PRACTICAL METHODS FOR ANALYSIS AND DESIGN OF HV INSTALLATION GROUNDING SYSTEMS

PRACTICAL METHODS FOR ANALYSIS AND DESIGN OF HV INSTALLATION GROUNDING SYSTEMS

LJUBIVOJE M. POPOVIĆ

ELSEVIER

ACADEMIC PRESS
An imprint of Elsevier

Academic Press is an imprint of Elsevier
125 London Wall, London EC2Y 5AS, United Kingdom
525 B Street, Suite 1800, San Diego, CA 92101-4495, United States
50 Hampshire Street, 5th Floor, Cambridge, MA 02139, United States
The Boulevard, Langford Lane, Kidlington, Oxford OX5 1GB, United Kingdom

Notices
Knowledge and best practice in this field are constantly changing. As new research and experience broaden our understanding, changes in research methods, professional practices, or medical treatment may become necessary.

Practitioners and researchers must always rely on their own experience and knowledge in evaluating and using any information, methods, compounds, or experiments described herein. In using such information or methods they should be mindful of their own safety and the safety of others, including parties for whom they have a professional responsibility.

To the fullest extent of the law, neither the Publisher nor the authors, contributors, or editors, assume any liability for any injury and/or damage to persons or property as a matter of products liability, negligence or otherwise, or from any use or operation of any methods, products, instructions, or ideas contained in the material herein.

Library of Congress Cataloging-in-Publication Data
A catalog record for this book is available from the Library of Congress

British Library Cataloguing-in-Publication Data
A catalogue record for this book is available from the British Library

ISBN: 978-0-12-814460-2

Publisher: Joe Hayton
Acquisition Editor: Lisa Reading
Editorial Project Manager: Carla Lima / John Leonard
Production Project Manager: Sruthi Satheesh
Cover Designer: Victoria Pearson

Typeset by TNQ Books and Journals

This book is dedicated to my grandchildren:
***Nadja, Andrej, Matija, and Maša**, wishing them to*
become good and useful people when they grow up.

CONTENTS

PREFACE

This book has appeared as a result of Author's intention for his research work in the field of substation grounding, which is a methodological and thematic whole to be presented as a scientific monograph. Also, he hopes that this book will enrich the existing professional and scientific literature necessary for all those dealing, or will be dealing, with this field of contemporary electrical engineering.

His research work, with a higher or lower intensity, lasted for nearly 30 years giving as the final result, in addition to a number of practical realizations, about 50 of research papers published in respectable international scientific journals or presented at international scientific or professional conferences.

The Author started his research activity by developing the system of equations representing general solution for discrete parameters ladder circuit. Later, he used these equations as theoretical foundation for his further investigations resulting in the analytical methods for solving a relatively wide spectrum of different grounding problems. Among them are those representing the most complex problems that can be encountered in practice, like the grounding problems of high voltage (HV) substations located in urban area. In the case of these substations the grounding system is spontaneously formed over a wide urban or suburban area surrounding these substations. The method presented here considers this type of grounding systems as an assemblage of several characteristic components, such as substation grounding electrode, outgoing cable lines acting as long grounding conductors, and/or grounding connections with the grounding systems of nearby medium voltage substations.

The book also presents the only so far known method, which enables determination of actual ground fault current distribution for a fault at an HV substation located in urban or suburban environment. By applying this method it is possible to take into account the influence of all, known and unknown, metal installations placed along and around HV lines passing through urban or suburban areas and put the whole grounding problem of supplied substations within a realistic framework. The results of experimental investigation carried out in the 110 kV distribution network of Beograd showed that, owing to the degree of urbanization, all potentials appearing in the grounding systems were several (three to five) times lower than it was considered before. This, of course, dramatically changes our perception of the whole grounding problem of HV substations located in urban or suburban areas. The possibilities for its efficient solution turned out much more favorable than it was considered before.

The presented methods, including their development, allow obtaining a good insight into the significance of individual relevant factors and parameters that certainly leads to a better understanding of the essence of the problem. The book also points out at all approximations and idealization of the actual physical model used in developing the presented analytical methods.

By applying the presented methods one can solve the problems, which are, owing to their complexity, not solvable by applying the available national, IEC, or IEEE standards. In that sense, this book can be considered as an enlargement of the corresponding technical standards, which enables obtaining solution of the most complex practical problems by using relatively simple calculation procedures. Then, it is interesting to say that the presented methods are more convenient than existing numerical methods applied in solving the most complex grounding systems of HV substations located in urban areas. The higher accuracy of the numerical methods usually cannot be utilized because of practical difficulties and impossibilities in collecting many relevant data about the metal installations located around these substations along and around their feeding lines.

The book is intended, first of all, for the professionals involved in the design of HV substations, but having in mind that the presented analytical methods give a good insight into interplay of essential parameters, it has much wider significance. Thus, it can be of interest for the academic people: professors, postgraduates, graduate, and undergraduate students of electrical engineering.

This book presents a description of the rounded up research results by the author mainly during his work in the Development Department of electric power distribution company "Elektrodistribucija Beograd," Serbia.

Ljubivoje M. Popović

CHAPTER 1

Introductory Considerations

1.1 GROUNDING IN GENERAL

From the very beginning of massive electrification and the widespread use of electrical energy, i.e., since the first decade of the last century, one of the most serious problems set before engineers/designers was how to connect electrical equipment and installations to the ground safely and economically. The problem of grounding has appeared in almost all areas of electrical engineering, from low-voltage stationary equipment and apparatus to electrical installations of high-installed powers. Because of ever-increasing power consumption and the associated developments of power systems, short-circuit currents have become increasingly dangerous and harmful, so a properly designed grounding system appears as an imperative not only for people working in high-voltage (HV) substations, but also for those working on nearby metal installations, such as water pipes, gas installations, telecommunication lines, railways, etc.

The two most important reasons why the grounding problem at the design stage of an electric power installation deserves a serious approach are because it:

- provides protection against electric shocks for the people inside, or in the vicinity, of the designed electric power installation in the case of a ground fault, and
- conducts a ground fault current into the surrounding earth without any undesirable consequences affecting the safe and reliable operation of the whole power system.

When a ground fault occurs in an electric power network, the earth as a specific conductive medium inevitably becomes a constituent part of the accidentally formed electric circuit. It acts as one of the available return paths of the ground fault current to its sources within the electric power system. However, from the standpoint of its physical appearance the earth as a conductive medium represents a three-dimensional body. Thus the freely involved space and mode of current flow through the earth is not simple, as it is in the case of well-known linear conductors used for constructing electric circuits for transmission or distribution of electrical energy.

Practical Methods for Analysis and Design of HV Installation Grounding Systems
ISBN 978-0-12-814460-2
https://doi.org/10.1016/B978-0-12-814460-2.00001-5

In selecting the minimum resistance opposing its flow, the current is dissipated with different intensities in all directions, creating widespread potential fields at the points of sinking to earth or coming to its surface. Certainly, under an unrealistic assumption that the earth is a conductive material, i.e., some kind of metal, the mentioned potential fields would not appear and the whole problem of electric power installation grounding would not matter at all. However, this is not the case in practice.

The electric conductivities of materials prevailing in the structure of the earth are very low compared to those of metals. Two basic constituents of earth, silicon and aluminum oxides, have electrical properties of insulators and the conductivity of the earth is due to the salts and humidity embedded between these two insulators. Although a relatively poor conductor by its structure, the earth is capable of conducting very high currents. Since the cross-section of the soil involved by current flows is freely formed, in view of the enormous dimensions of the earth it can enlarge practically with no limits. Only at the points of sinking or coming to the earth's surface is this cross-section limited to the contact surface between the grounding electrode and the surrounding earth. This is the reason why along the entire current path through the earth at these places the resistances to its flow are the highest. These resistances represent the grounding electrode resistances. For a given value of the surrounding soil resistivity and the shape and spatial position of a grounding electrode, the value of this resistance depends only on the magnitude of the electrode. Thus when we desire a lower value of this resistance in a concrete case, we must enlarge the dimensions of the considered grounding electrode. The formulas for calculating this resistance of grounding electrodes of different magnitude, geometric shapes, and spatial positions are usually given in the corresponding technical standards.

The forms of equipotential contours of the electric field in the vicinity of an HV substation closely follow the external shape of its grounding electrode. Further away, their forms change and after a certain distance, if conductivity of the soil is homogeneous, their form becomes spherical, irrespective of the size and form of the considered grounding electrode. Since people are normally situated on the surface of the soil, for solving practical problems of significance are only the equipotential curves, i.e., the potential distribution, at the earth's surface and the potentials of grounded metal constructions that could become the subjects of human touch. Practically, this means that by selecting an adequate magnitude and shape of the substation ground grid, we can obtain the most favorable surface potential distribution for a priori known equipment disposition. As an illustration, the potential distribution on the surface of the soil of a typical ground grid laid in a homogenous soil is shown in Fig. 1.1.

The notation used has the following meaning:

- U_t and U_s—touch and step voltages, and
- V_g—potential of the substation ground grid, or ground potential rise [1].

From Fig. 1.1 one can obtain an idea of the influence of space position of individual elements of the grid upon potential distribution on the surface of the soil.

Calculations of the potential distribution on the surface of the soil, above and in the vicinity of a ground grid, are very complicated and extensive, even if it is assumed that the soil is homogeneous and the ground grid is of a geometrically uniform shape, i.e., the grounding conductors forming the grid are uniformly spatially disposed.

For the purpose of estimation of the achieved safety conditions, it is necessary to determine the value of potential V_g first. If its value is smaller than the maximal tolerable touch voltage, then the required safety conditions are obviously achieved (Fig. 1.1). However, if this criterion is not satisfied and the value of V_g is higher than the maximal tolerable touch

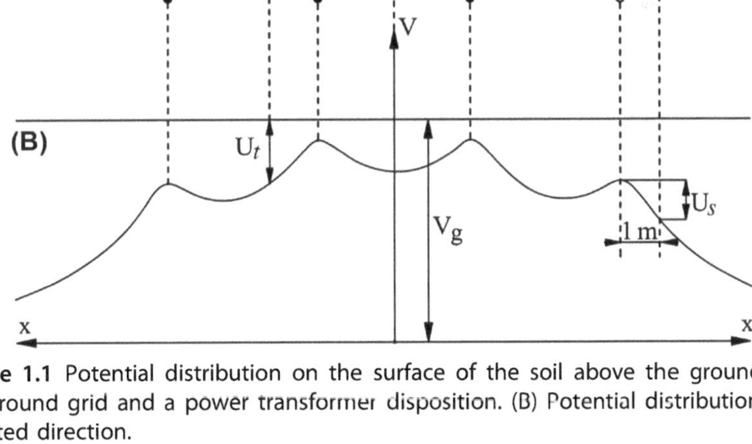

Figure 1.1 Potential distribution on the surface of the soil above the ground grid. (A) Ground grid and a power transformer disposition. (B) Potential distribution for a selected direction.

voltage, in such a case the maximal calculated touch and step voltages become relevant for estimation of the achieved safety conditions. For a grounding grid of certain form and size the critical potential differences (the maximal touch and step voltages) can be estimated by using the simplified calculation procedure usually given in the corresponding technical standards, e.g., Ref. [1].

During a ground fault, the current flows and the corresponding raised potentials appear at places where they do not exist under normal operating conditions. Hazardous voltages appear as a consequence of excessive potential differences between points at such distances that can be simultaneously bridged by the human body. These are caused by potential gradients at various directions along the surface of the soil and potential differences between grounded metal structures and the surrounding soil (Fig. 1.1). For the purpose of establishing unique criteria for the estimation of safety conditions, these potential differences have been standardized by introducing precisely defined terms of the touch and step voltages. At the design stage of an electric power installation, it is required to show by calculations that these voltages, at any place within or in the vicinity of this installation, under conditions of the worst (critical) ground fault, will stay within the prescribed limits. The maximum permissible values of the touch and step voltages are dependent on the time duration of the current exposure and are strictly determined by the corresponding technical standards/guidelines, [1].

The problem of estimation of safety conditions in the case of substations operating in medium voltage (MV) networks having neutral point(s) isolated or grounded through a resistor/reactor, i.e., where the ground fault current that is restricted to an a priori adopted and relatively small value, is usually solvable without any special analysis. The restricted value of the ground fault current allows the problem, in the majority of cases, to be solved simply by applying a standardized solution from the standpoint of the shapes and dimensions of the designed substation grounding electrodes. Nevertheless, under exceptionally unfavorable conditions, e.g., high soil resistivity, the need may arise to perform a grounding system analysis with the aim of finding a specific solution acceptable only in the considered concrete case.

1.2 GROUNDING SYSTEMS IN GENERAL

HV networks have directly, or through low resistance, grounded neutral points and as a result the ground fault currents in HV substations may reach more than 20 kA. On the other hand, the neutral conductors of the connected transmission and/or distribution lines (overhead ground wires and/or cable metal sheaths), as is well known, act as specific long grounding conductors and improve characteristics of the grounding system of these substations. Thus the grounding problem of these substations is quite different in comparison with MV/low-voltage (LV) distribution substations.

There are many factors influencing the values of the touch and step voltages at certain places within or in the vicinity of an HV substation. The most important are: local soil resistivity, ground fault current, dimensions and form of the applied grounding electrode, and number and type of the connected transmission and/or distribution lines. Each has a specific and large influence in creating substation ground grid potential and potential differences between different points on the earth's surface (Fig. 1.1). As a result they should be taken into account by designer calculations as accurately as possible.

A practical grounding system is usually composed of the grounding grid of the substation itself and many conductively connected neutral conductors: ground wires of overhead lines and metal sheaths of cable lines. In addition to these neutral conductors, the grounding system can sometimes involve different metal pipelines, foundation armatures of nearby buildings, etc. Thus it can be said that the grounding system of an HV substation represents an assemblage of mutually different elements from the standpoint of their basic function. Since all of the mentioned metal installations exist for other reasons, i.e., independently of our need to solve the grounding problem, it is quite reasonable to take their grounding contribution into consideration. Since the technical characteristics, mutual disposition, and number of these installations are determined in accordance with their primary function, we only need to take into account their grounding contributions as accurately as possible. Along these lines we place the whole grounding problem of an HV substation in a realistic framework leading to a solution that is in accordance with real needs. However, for this it is necessary to have a corresponding mathematical model to calculate the grounding system characteristics by taking into account all relevant factors and parameters. Since relevant factors and parameters are different for each case it is not possible to define a ground electrode that would represent a universal (standardized) solution applicable for each concrete HV substation.

The source or transmission substations having connected overhead lines only are relatively easy for grounding system analysis and designing a corresponding ground grid [1,2]. The simplicity of the necessary calculation procedure emerges because the substation ground grid and transmission tower footing electrodes are mutually so distant from each other that there is no interference in dissipating ground fault current into the surrounding earth.

In the case of a grounding system belonging to an HV distribution substation located in an urban or suburban area the situation is much more complex. The urban area encompassed by the grounding system of these substations is much larger than the area covered by the substations themselves. Because of the large number of known and unknown metal installations laid in different directions and frequently in an enclosed space along city streets, the grounding systems spontaneously formed in typically urban surroundings are the most complex that can be encountered in practice. On the other hand, because of the large area encompassed by their external electrodes, mainly metal sheaths of outgoing MV cable lines, grounding systems of HV distribution substations have excellent characteristics, including a very low ground impedance value. Thus as a rule the problem appears only at the design stage, when the grounding contribution of the many outgoing cable lines acting as long grounding conductors and/or conductive connections with the grounding systems of supplied MV/LV substations should be taken into consideration.

Contemporary methods for determining the relevant characteristics of grounding systems in various practical cases have been developed as the result of a lengthy evolution in the search for the safest and most economical solutions. An enormous number of scientific and professional papers were published by many authors who have contributed to the present level of knowledge in this area. Unfortunately, here it is possible to mention only a small number of them.

The complexity of the problem and difficulties caused by deficiency or uncertainty of relevant data have had a significant influence on the manner of the development of necessary calculation methods. As a consequence, two different tendencies can be identified. On the one hand, research efforts have been made to develop a mathematical model that takes into account, as accurately as possible, many relevant factors and parameters, including those of lower significance, e.g., Ref. [3]. On the other hand, by applying reasonable assumptions and idealizations to the real physical model, efforts have been made to develop directly applicable and relatively simple methods that are sufficiently accurate for practical purposes [4].

From the standpoint of the relevant data that can be taken into account, methods of high capability have appeared with the development of computer technology and the progress made in the numerical analysis of electromagnetic fields. In principle, they can be used for solving complex practical problems, such as grounding systems of HV substations located in an urban area. However, difficulties associated with the application of these methods have already appeared with the collection of input data necessary for obtaining the final results of high accuracy. Due to the difficulty of collecting all relevant/input data of local soil and metal installations surrounding an HV distribution substation and its feeding line, higher accuracy associated with the application of numerical methods cannot be practically utilized.

Practical methods are mainly based on analytical procedures accompanied by corresponding reasonable assumptions and approximations, which are usually conservative from the standpoint of possible undesirable consequences. By their application, one obtains a relatively clear picture of mutual relations and significance of individual factors and parameters. Insight into the influence of important factors is not obscured by a large number of factors of low order, as is the case with numerical methods. This certainly contributes to an easier and better understanding of the essence of the problem and makes these methods more desirable for practical application. Thus, from the point of view of practical application, numerical methods should be considered necessary mainly for verifying the applicability of corresponding analytical methods.

By considering existing methods, one more aspect of the nature of the whole problem should be taken into account. The most important goals of substation grounding are saving human lives and sensitive electronic equipment without causing any damage because of a possible ground fault. This clearly indicates that more desirable methods for performing necessary calculations are not those that give results of high accuracy, but those that give results containing certain safety margins with respect to undesirable consequences.

The method presented here represents the only known analytical method for performing the analysis of complex grounding systems spontaneously formed in urban or suburban areas [4,5]. It has been developed under the following assumptions and approximations. The grounding contribution of the substation grounding electrode/grid is calculated as if it were the only existing component of the whole grounding system. Then, the grounding contribution of the individual cable line metal sheath(s) is calculated depending on the space disposition in relation to the substation grounding electrode and to the other cable lines. Accordingly, all outgoing cable lines are divided into two characteristic groups: cable lines that, over their entire active lengths, are laid in one of usually four primary directions and cable lines that are diverted at some distance, less than their active length, into a different direction, commonly called a secondary direction. The grounding contribution of a cable line belonging to the first group is calculated by using an analytical expression that takes into account mutual interference between this cable line and the substation grounding electrode in dissipating the ground fault current into the surrounding earth. The grounding contribution of a cable line belonging to the second group is calculated by using an analytical expression that disregards this cable line's contribution along the section where it is laid in the same trench with other cable lines. In this way the complicated calculations concerning the influence of the mutual interference between this cable line and all other cable lines in dissipating fault current into the earth are essentially simplified. Also the mutually inductive coupling between this cable line and all other cable lines in transferring the ground fault current through the cable's metal sheaths is avoided. The assumptions and approximations are such that the application of this method gives results that are slightly on the conservative side. However, determination of the grounding contribution of all external grounding electrodes is insufficient to correct the final design and construction of the ground grid of an HV substation. This element of the grounding system is designed only for grounding purposes and should be adjusted through the designing procedure for real needs, i.e., without excessive expenditure regarding materials and associated labor. To achieve this goal, one more aspect of the whole problem should be considered and solved.

During the ground fault in a power system, the fault current leaves the faulted phase conductor and returns to the generating sources in a power system through the earth and neutral conductor(s) of the feeding line, ground wire(s) of the overhead line, and metal sheaths of the cable line. Thus the problem of determining the ground fault current distribution, i.e., determining the part of the ground fault current dissipated into the surrounding earth through the substation grounding system, should also be solved. In the case of a generating (power or transmission) station this means that the part of the ground fault current appearing in the transmission line ground wire(s), as a consequence of its inductive coupling with the phase conductors, has to be determined. The induced current reduces ground fault current passing through the substation grounding system and the surrounding earth in comparison with an unreal assumption of the same line, but without ground wire(s).

The problem of developing a method for determining the influence of the earth as a return path on line conductor impedances was investigated during the 1930s, whereas investigation of the problem of determining the ground fault current distribution has lasted far longer. The first methods developed for solving this problem were related to the case where the feeding line was constructed as an overhead line. Later, scientific papers considering the problem in special cases where a feeding line appeared as a longitudinal combination of one cable and one overhead section were also published. Since then, several methods have been developed for solving this problem in cases where, because of a high local soil resistivity and/or a high short-circuit level, special measures aimed at reducing a fraction of the ground fault current dissipating into the earth (e.g., counterpoises, bare copper conductor laid in the same trench as the cable feeding line, etc.) were considered indispensable. Also more complicated problems appeared in special cases where many ground faults in the outgoing overhead lines for the purpose of determining critical fault position had to be analyzed. Finally, a method developed for determining ground fault current distribution when a feeding cable line is constructed of three single-core cables was presented. The method took into account participation of all three metal sheaths in the ground fault current distribution for a fault anywhere along the feeding line [6,7].

The common characteristic of these methods is that they take into consideration only the design/constructive characteristics of the feeding line and the main characteristic of the surrounding earth as a conductive medium. However, in the case of HV distribution substations located in urban or suburban areas, it is not possible to obtain the actual value of the part of the ground fault current dissipated into the surrounding earth by using the formerly mentioned methods. Feeding lines of these substations pass through urban areas and many metal installations are located along and around them. Their structures and topologies are usually very complex and depend on many local factors. The ground fault current portion returning through this spontaneously formed underground metal network reduces the current passing through the earth. This is certainly a very favorable effect, but a problem arises as to when it should be evaluated. Because both of the mentioned fault current portions pass through the substation grounding system, they are separated along many of its elements. Thus there is no way of obtaining the value of any of these two currents purely by calculations for the following reasons:

- The actual physical model (electrical circuit) is very complex.
- The fault current is dissipated into the earth with different intensities along many different grounding system elements.
- The data of many of the underground metal installations surrounding the feeding line are partially or completely unknown.

Thus none of the previously mentioned methods provides the correct solution to the problem when it appears in urban or suburban conditions, i.e., when many other metal installations participate in the ground fault current returning to the power system. In such conditions, each power line becomes a very complex electrical circuit with many conductively and inductively coupled elements having uncertain or completely unknown relevant parameters. As a result the problem concerning ground fault current distribution, when an HV line supplies a substation located in an urban or suburban area, has lacked an adequate solution for a long time. The practical problems were, however, solved by a rough pessimistic approximation that surrounding metal installations did not exist.

1.3 FAULT CURRENTS PASSING THROUGH GROUNDING SYSTEMS IN URBAN AREAS

This problem has been known for a long time and in professional and scientific circles, within the framework of electrical and electronic engineering, one needs to ask: how can one determine mutual, inductive influence between an electric power line and metal installations surrounding this line in urban and/or suburban areas? In other words, from an electrical circuit theory point of view, this question can be put another way: how can one determine the electrical characteristics of many inductively coupled conductors having the earth as common return path and many unknown but relevant data?

Since along and around each electric power line a fluctuating magnetic field is an accompanying phenomenon, it is not difficult to conclude that this problem appeared in practice simultaneously with the first applications of Tesla's alternative currents, i.e., at the outset of the 20th century. In principle, the problem becomes solvable after the appearance of Carson's theory of current return path through the earth, i.e., the development of analytical expressions for calculating the self- and mutual impedances of inductively coupled circuits in the second half of the 1930s.

Usage of common routes (mainly sidewalks) for various supply networks (electricity, water, gas, oil, telecommunications, etc.) unavoidably leads to the appearance of the mutually inductive influence of these installations. Therefore each power line passing through urban and/or suburban areas is without exception inductively coupled with many other metal installations typical for urban areas. However, the relevant parameters of these lines used to be determined by completely ignoring the existence of surrounding metal installations.

It was known that surrounding metal installations have a favorable influence on the transfer characteristics of power lines and on the grounding problem of supplied substations by reducing part of the ground fault current returning to the power system only through the earth. However, until recently, none of the known, and practically applicable, methods for performing necessary calculations took into consideration the existence of surrounding metal installations. As a result the following question can be posed: why has the problem that has existed from the very beginning of the application of Tesla's alternative currents remained so long without an adequate solution?

The structure and topology of a large network, composed of all the metal installations of any contemporary town, are almost completely unknown because they are situated mainly belowground. Thus many relevant data about them cannot be visually or in any other way determined or verified. In other words, since many urban metal installations are invisible, necessary data such as constructive characteristics and the mutual space dispositions of these installations, including in some cases even their number along the considered power line, are uncertain or completely unknown. Also the surrounding metal installations impede the attainment correct measurement results of specific soil resistivity, and therefore the equivalent resistivity of the surrounding soil, a basic piece of data, also remains an unknown but relevant parameter. Because of all of these practical difficulties, the planners, designers, and management of a distribution network and its lines were forced to disregard completely the existence of surrounding metal installations when calculating necessary impedances of these lines.

These practical difficulties are also the reason why measurements of the part of ground fault current returning to the power system only through the earth are impossible to get. This is because in urban conditions another fraction of ground fault current returns to the power system through surrounding metal installations and cannot be determined by calculations or by measurements. Both of these two ground fault current fractions pass through the grounding system of the supplied substation and because they are mutually separated they continue separately their flows belowground. Therefore, for this purpose, none of them can be measured at some accessible and suitable location. The process of their mutual splitting and the dissipation of one of them into the surrounding earth occur continuously along numerous and very long external electrodes belonging to the grounding system of each substation located in the urban surroundings [4,5].

However, experimental investigations performed about 10 years ago with 110 kV cable lines in the distribution network of Beograd, Serbia, showed that the previously described difficulties can be overcome [7]. The developed method is based on measurement of the cumulative effects of the surrounding metal installations upon the value of the current induced in one of the metal sheaths (neutral conductors) of the considered feeding cable line. The solution is achieved by introducing into the developed analytical procedure a cylindrically shaped fictitious conductor around and along the feeding cable line. Other necessary parameters concerning this conductor are determined under the condition that its inductive influence upon the ground fault current distribution is identical to the inductive influence of all, known and unknown, metal installations realy existing along and around a considered cable line.

Furthermore, the achieved solution has now been extended to include overhead distribution lines by defining the location along these lines where the current induced in the ground wire(s) should be determined by measurements [8—11]. In this way the method for determining the ground fault current fraction induced in all metal installations surrounding a feeding line, as well as the actual ground fault current fraction that, in urban environments, returns through the earth, is completed. The analytical procedure for determining all relevant parameters of the mentioned fictitious cylindrical conductor for any distribution line has also been developed.

The results obtained by these investigations show that the ground fault current fraction dissipating into the surrounding earth in a typical urban environment is 3—5 times smaller than has been considered before. This fact throws new light on the grounding problem of urban HV/MV substations and dramatically changes the former perception of the magnitude of this problem. Now, it can be seen within a realistic framework that leads to less conservative or more economical solutions in each concrete case. Greater economic effects can certainly be expected in cases where, because of high soil resistivity and/or high short-circuit current level, special measurements (e.g., bare copper conductor laid in the same trench as the cable feeding line, counterpoises, etc.) were foreseen. In addition, one can expect removal of the strict requirement for application of expensive MV cables acting as grounding conductors (cables with an uncovered metal sheath) as was the case with the MV network of Beograd. A small difficulty arises, because the actual ground fault current distribution depends on the urban metal installations surrounding the considered feeding line, i.e., local conditions different for each concrete case. This means that to obtain ground fault current distribution (without ignoring the surrounding metal installations), or a feeding line actual reduction factor, each concrete HV distribution line should be treated separately from the corresponding measurements and calculations. As is well known, this was not the case in the past, because cable line manufacturers' data regarding their reduction factor value were used, and just after the method was developed in Refs. [6,12] the reduction factor of cable lines consisting of single-core cables was determined by designer calculations.

Thus the previously mentioned method compensates for the objective deficiency of an unlimited number of relevant data concerning the surrounding soil and metal installations through the field measurement of their cumulative effect upon the value of the test currents in phase conductors and in one of neutral conductors of the considered HV line. This is based on the results of on-the-spot measurement results and the well-known analytical expression for self- and mutual impedances of line conductors in the semiinfinite and homogeneous earth [2].

The whole problem of the inductive influence of the surrounding metal installations has four important aspects i.e. can be considered from the standpoint of:

- the inductive influence of the surrounding metal installations on the power line transfer characteristics during normal operation;
- the inductive influence of the surrounding metal installations on the power line characteristics during fault conditions;
- the dangerous and harmful inductive influence of a power line on surrounding metal installations during normal operation; and
- the dangerous and harmful inductive influence of a power line on surrounding metal installations during fault conditions.

Thus the formerly depicted method with certain small modifications can be used for the following problems in current electrical engineering practice:

- Determination of the actual transfer characteristics of power lines passing through urban and/or suburban areas;

- Determination and elimination of the dangerous and harmful influence of power lines on any of the urban metal installations during normal operation; and
- Determination and elimination of signal disturbances in the surrounding communication circuits.

As has already been mentioned, the introduced fictitious conductor has the following physical appearance and disposition: it is of cylindrical shape and surrounds all feeding line conductors over its entire length. In accordance with this, it is clear that this conductor acts as an additional neutral conductor for each line passing through urban and/or suburban areas. This additional neutral conductor certainly improves the transfer characteristics of these power lines. Thus when all relevant parameters of the introduced fictitious conductor are determined for a certain power line, its actual transfer characteristics can be obtained by using a well-known calculation procedure for the determination of power line series impedance and by treating this conductor as an additional neutral conductor of the considered line [13].

In the case of determining the inductive influence of a power line upon a separate metal installation exposed along the whole length of this power line, the equivalent neutral conductor can be determined as having an inductive effect, substituting the inductive effects of all realy existing surrounding metal installations, including neutral conductors of the considered power line [9]. When such an installation is exposed only along one section of the considered power line, this equivalent conductor is also of cylindrical shape, but then involves the considered metal installation only along the section of exposure, acting as an equivalent screening conductor [14].

1.4 THE CORRESPONDING STANDARDS AND THE METHODS PRESENTED HERE

1.4.1 A Short Overview of the Existing International Standards

The aim of unification of criteria for all HV substations is to define fundamental requirements for safety through corresponding national and international technical standards. Since the criteria defined through the individual national standards are somewhat different from country to country we will here consider only the criteria defined by international standard [1].

The specific purposes of this standard (guide) are to:

- establish, as a basis for design, the safe limits of potential differences that can exist in a substation under fault conditions between parts that can be contacted by the human body;
- review substation grounding practices with special reference to safety, and develop criteria for a safe design;
- provide a procedure for the design of a practical grounding system, based on these criteria.

According to this standard, a safety ground design has the following two objectives:

- To provide a means to carry electric currents into the earth under normal and fault conditions without exceeding any operating and equipment limits affecting continuity of service;
- To ensure that a person in the vicinity of the grounded facilities is not exposed to the danger of critical electric shock.

Also this standard defines a practical approach to safe grounding concerns and services, as follows:

- The intentional ground, consisting of ground electrodes buried at some depth below the earth's surface;
- The accidental ground, temporarily established by a person exposed to potential gradient in the vicinity of the ground facility (Fig. 1.1).

If the magnitude and shape of ground electrodes, local soil resistivity, and other factors produce an excessive potential gradient, the grounding system may be inadequate despite its capacity to carry the fault current in magnitude.

With the aim of establishing the unique criteria for numerous practically possible conditions and circumstances of electric shock, the existing technical standards introduce the precisely defined terms of step and touch voltage. They are defined as follows:

- The potential difference between the ground potential rise and the surface potential at the point where a person is standing while at the same time having a hand in contact with grounded structure (Fig. 1.1);
- The difference in surface potential expressed by a person bridging a distance of 1 m with the feet without contacting any grounded object (Fig. 1.1).

For estimation of the safety condition in a gas-insulated substation it was necessary to introduce and define another touch voltage. This is metal-to-metal touch voltage.

The most common physiological effects of an electric circuit on the body, in order of increasing current magnitude, are: threshold perception, muscular contraction, fibrillation of the heart, respiratory nerve blockade, and burning. If the shock currents are kept below the fibrillation threshold by a properly designed grounding system, injury or death may be avoided.

As shown by Dalziel and others, the nonfibrillation current of magnitude I_b at duration, ranging from 0.03 to 3.0 s related to the energy absorbed by a body, is described by the following relation:

$$S_b = (I_b)^2 t_s \tag{1.1}$$

where I_b is the root mean square magnitude of current through the body (A), t_s is the duration of exposure to the current (s), and S_b is the empirical constant related to the electric shock energy tolerated by a certain percentage of the population.

The internal resistance of the body is approximately 300 Ω, whereas values of body resistance, including the skin, range from 500 to 3000 Ω, as suggested by Delziel and others. By using this knowledge as a standard [1] introduces the following safety criteria:

- For body weight 50 kg:

$$E_{sp} = (1000 + 6C_S r_S \rho_S)\frac{0.116}{\sqrt{t}}, \text{(V)} \tag{1.2}$$

$$E_{tp} = (1000 + 1.5C_S r_S \rho_S)\frac{0.116}{\sqrt{t}}, \text{(V)} \tag{1.3}$$

- For body weight 70 kg:

$$E_{sp} = (1000 + 6C_S r_S \rho_S)\frac{0.157}{\sqrt{t}}, \text{(V)} \tag{1.4}$$

$$E_{tp} = (1000 + 1.5C_S r_S \rho_S)\frac{0.157}{\sqrt{t}}, \text{(V)} \tag{1.5}$$

where E_{sp} is the maximal permissible step voltage (V), E_{tp} is the maximal permissible touch voltage (V), r_S is the resistivity of the surface material (Ωm), and t is duration of the shock current.

The coefficient C_S represents a corrective factor to compute the effective foot resistance in the presence of surface material and can be determined by:

$$C_S = 1 - \frac{0.09\left(1 - \dfrac{\rho}{\rho_S}\right)}{2h_S + 0.09}, \tag{1.6}$$

where ρ_S is the surface material resistivity (Ωm), ρ is the resistivity of the earth beneath the surface material (Ωm), and h_S is the thickness of the surface material (m).

If no protective surface layer is used, then $C_S = 1$ and $\rho_S = \rho$.

With the substation of $\rho_S = 0$ in the foot resistance terms of Eqs. (1.3) and (1.5), we have metal-to-metal touch voltage, and the maximal permissible voltage is determined by:

- For body weight 50 kg:

$$E_{mm} = \frac{116}{\sqrt{t}}, (V) \tag{1.7}$$

- For body weight 70 kg:

$$E_{mm} = \frac{157}{\sqrt{t}}, (V) \tag{1.8}$$

where E_{mm} is the maximal permissible metal-to-metal touch voltage (V).

For the purpose of efficient solving of the problem, another international technical standard important for HV substation grounding has been established by [2]. This standard specifies procedures for calculation of the prospective short-circuit currents with an unbalanced short circuit in an HV three-phase AC system operating in nominal frequency 50 or 60 Hz, i.e.:

- currents during two separate simultaneous line-to-earth short circuits in insulated or resonant earthed neutral systems, and
- partial short-circuit currents flowing through earth in the case of a single line-to-earth short circuit in a solidly earthed or low-impedance earthed neutral system.

The currents calculated by these standardized procedures are used when determining induced voltages or touch or step voltages and a rise of earth potential at an HV substation and the towers of overhead HV lines as well as partial short-circuit currents through the earth, ground wires of overhead lines, and sheaths or shields of cables leading to conservative results with sufficient accuracy.

1.4.2 The Methods Presented Here in Comparison With Existing International Standards

This book presents the development of the analytical methods that enable the solving of complex grounding problems of HV substations by using relatively simple calculation procedures. Some of these problems cannot be solved by using the existing technical standards [1,2]. From that point of view this book can be considered as an enlargement of the area covered by the corresponding international and a large number of national standards. In brief the contents of this book are as follows:

- Chapter 1 presents the introductory and general considerations concerning the grounding and grounding systems of HV substations. Also this chapter gives a brief overview of the historical development and current methods available for analysis and design of grounding systems of HV substations. This overview, inter alia, involves a brief comment on the methods presented in the book in comparison with the methods available in the corresponding international standards. The main conclusion is: the methods presented here enlarge the application field in comparison with the methods offered through contemporary international standards.
- Chapter 2 presents the derivation procedure of the general equations of lumped parameter ladder circuits and their application in the analysis of the relevant characteristics of long grounding conductors such as ground wires of overhead lines and metal sheaths of cable lines. Also these equations are used for the derivation of the analytical expression for estimation of the active length of different metal installations acting as long grounding conductors.
- Chapter 3 presents the measurement procedure that, together with the corresponding analytical procedure, represents the only so far known method enabling determination of the actual reduction factor for HV lines passing through urban and/or suburban areas. It takes into consideration the contribution of all known and unknown metal installations typical in urban areas to the reduction in ground fault current dissipated into the surrounding earth through the grounding system of a supplied HV substation. This certainly leads toward more economical solutions for the grounding problem of HV substations located in urban or suburban areas. The problem appears because many urban metal installations are situated belowground and can be neither visually nor in some other way determined or verified. Because of this, data on the metal installations surrounding a feeding HV line necessary for the determination of self- and mutual impedances of all conductors in the arrangement are uncertain or completely unknown. The developed method compensates for all data deficiencies concerning metal installations involved with the fluctuating magnet field appearing around and along a feeding line independent of their number, space disposition, and technical characteristics. From that point of view, this book can be considered as an enlargement of the application field of standard [1].

Also presented in Chapter 3 is the derivation of the analytical expression for the determination of ground fault current distribution for a fault at any place along any type of HV and MV cable line. This makes it possible to choose cable(s) for the feeding line with a metallic sheath that is in accordance with real needs, or large enough to withstand thermal stress in ground fault conditions, but not too large to cause expensive cable production or too high values of harmful circulating currents during normal operation conditions of the cable line. The finally obtained analytical expression has already been introduced in technical standard [2].

- Chapter 4 presents the analytical method for determining the feeding line reduction factor in specific practical cases, when the considered feeding line cannot be treated as homogeneous along its entire length, e.g., when it contains one overhead and one cable section. From that point of view this method can be considered as an extension of the application field of standard [1]. Also Chapter 4 presents an analytical method for determining the ground fault current distribution when special measures are foreseen with the aim of reducing the part of the fault current dissipated into the earth through the grounding system of a supplied HV substation. From that point of view this method can be considered as an extension of technical standard [1].
- Chapter 5 presents the analytical procedure for determining the critical (from the standpoint of the part of the ground fault current passing through the grounding system of a generating station) fault position for ground faults in outgoing HV overhead lines. The presented method is less approximate than the one presented in standard [2].
- Chapter 6 presents the only so far known analytical method for determining the relevant characteristics of large and complex grounding systems spontaneously formed around HV substations located in urban or suburban areas. The method presented here considers these types of grounding systems as an assemblage of several characteristic components, such as substation grounding electrode, outgoing cable lines acting as long grounding conductors, and/or conductive connections with the grounding systems of supplied MV substations. By this, their mutual interference in dissipating a ground fault current into the surrounding earth is taken into account by relatively simple analytical expressions, or without necessarily relying on modern computers. The method solves the problem in each practically possible case by using a relatively simple calculation procedure. From that point of

view this method can be considered as an extension of the application field of standard [1].

- Chapter 7 presents the measurement procedure, which together with the corresponding analytical procedure, represents the only so far known method enabling determination of the voltages and currents induced by a ground fault current in an HV line on any of the nearby metal installations, seen separately among many other known and unknown metal installations present in urban surroundings. The problem emanates from the fact that the ground fault current leaves the power system and in returning to the their sources uses all available paths, including those that are not foreseen for this purpose. This means that dangerous and harmful currents and voltages can appear at places that are very distant from the HV substation in which the ground fault occurs.
- Chapter 8 presents the only so far known testing procedure enabling correct estimation of grounding systems of HV substations located in urban and suburban areas by elimination of the unavoidable influence of metal installations surrounding a test circuit upon the obtained results of measurements. The presented research results show that ignoring the existence of surrounding metal installations can lead to a completely incorrect estimation of safety conditions achieved in the tested HV substation.

The methods presented in Chapters 3 and 6 represent one complete analytical procedure for determining the relevant characteristics of large and complex grounding systems spontaneously formed around HV distribution substations located in urban or suburban areas. It solves the problem in each practically possible case by using a relatively simple measurement procedure and mathematical operations. In that sense, this procedure can be considered as an extension of the application field of existing standards [1,2].

The experience gained so far shows that the contribution of the outgoing cable lines as long grounding conductors and metal installations surrounding feeding lines as the returning paths is essentially important. In the case of 110/10 kV gas-insulated substations supplied by the cable distribution network of Beograd, it was noticeable that there was no need to lay into the earth any grounding conductors dedicated to grounding purposes only. Irrespective of the relatively high values of ground fault currents of the order of 20 kA, metal armatures of the foundations of the buildings of these substations were sufficient to serve as the substation grounding electrode. Certainly, such solutions are economically extremely favorable. All grounding functions are overtaken by metal installations that exist independently of the need to solve grounding problems at these substations.

Finally, since in this field approximate estimation and engineer's intuition play significant roles, perhaps greater than in any other field of modern electric power engineering, the book also includes several quantitative analyses and illustrative numerical examples.

REFERENCES

[1] IEEE Guide for Safety in Substations Grounding, Std. 80—2000.
[2] Short-circuits Currents in Three-phase A.C. System-part 3: Currents During Two Separate Simultaneous Line-to-earth Short Circuit and Partial Short-circuit Flowing Trough Earth, IEC Int. Std. 60909—3, Ed. 3.0, 2009—03.
[3] J. He, R. Zeng, B. Zhang, Methodology and Technology for Power System Grounding, WILEY and IEEE, 2014.
[4] Lj.M. Popović, Practical method for the analysis of earthling systems with long external electrodes, IEE Proceedings C 140 (3) (1993) 213—220.
[5] Lj.M. Popović, Comparative analysis of grounding systems formed by MV cable lines with either uninsulated or insulated metal sheath(s), Electric Power System Research 81 (2) (2011) 393—399.
[6] Lj.M. Popović, Determination of the reduction factor for feeding cable lines consisting of three single-core cables, IEEE Transactions on Power Delivery 18 (3) (2003) 736—743.
[7] Lj.M. Popović, Influence of the metallic installations surrounding the feeding cable line on the ground fault current distribution in supplied substations, IEEE Transactions on Power Delivery 23 (4) (2008) 2583—2590.
[8] Lj.M. Popović, Determination of the actual reduction factor of HV and MV cable line passing through urban and suburban area, FACTA UNIVERSITATIS-Series: Electronics and Energetics 27 (1) (2014) 25—39.
[9] Lj.M. Popović, Ground fault current distribution when a ground fault occurs in HV substations located in an urban areas, Progress in Electromagnetic Research B 59 (2014) 167—179.
[10] Lj.M. Popović, Transfer characteristics of electric-power lines passing through urban and suburban areas, International Journal of Electrical Power & Energy Systems 56 (2014) 151—158.
[11] Lj.M. Popović, Influence of surrounding metal installations on the parameters of electric-power lines passing through urban and suburban areas, Chapter in the book, in: Materials and Processes for Energy: Communicating Current Research and Technological Developments, Formatex Research Centre, 2013, pp. 831—839. Energy Book Series-2013 Edition, Spain.
[12] Lj.M. Popović, Improved analytical expressions for the determination of the reduction factor of the feeding line consisting of tree single-core cables, European Transactions on Electrical Power 18 (2) (2008) 190—203.
[13] Lj.M. Popović, Testing and evaluating grounding systems for substations located in urban areas, IET Generation Transmission and Distribution 5 (2) (2011) 231—238.
[14] Lj.M. Popović, New method for determination of inductive influence of one or more power lines on any of exposed metal installations, International Journal of Engineering Research and Application 6 (2) (2016) 36—46.

FURTHER READING

[1] Lj.M. Popović, Ground fault current distribution in a station supplied by a line composed of a cable an overhead section, European Transactions on Electrical Power 17 (3) (2007) 207–208.
[2] Lj.M. Popović, Efficient reduction of fault current through the grounding grid of substation supplied by cable line, IEEE Transactions on Power Delivery 15 (2) (2000) 556–561.
[3] Lj.M. Popović, Practical method for evaluating ground fault current distribution in station, towers and ground wire, IEEE Transactions on Power Delivery 13 (1) (1998) 123–129.
[4] CCITT Directives Concerning the Protection of Telecommunication Lines Again Harmful Effects from Electric and Electrified Railway Lines, ITU, Geneva, 1990 (project report).

CHAPTER 2

Theoretical Foundations

2.1 GENERAL EQUATIONS OF DISCRETE PARAMETER LADDER CIRCUITS

2.1.1 Introduction

It is well known that lumped parameter ladder circuits represent universal physical and mathematical models of systems having distributed parameters. They are useful in performing analyses of the behavior of electrical systems, including electronics, as well as analysis of heat transfer, acoustics, etc. Since in current engineering practice some of the ladder circuits have by their nature elements that are discrete, e.g., electrical circuit formed by transmission line ground wires, in such cases the lumped parameter ladder circuits are quite adequate as mathematical models.

The behavior of voltages and currents along a transmission line is described by linear partial differential equations, the celebrated telegrapher's equations. However, the derivation of these equations is only the introduction to a study of transmission lines; the important thing is their solution.

The existence of sinusoidal oscillations covering a large range of frequencies, and development of modulators allowing information to be placed on such a signal, has led to the complete development of the solution to the telegrapher's equations for sinusoidal time function, known as the general line equations, e.g., Ref. [1]. However, they are convenient only for analysis of the transmission line, which is in a steady-state regime. By applying them and associated other procedures, influences and mutual relationships between parameters relevant to resonant phenomena as well as transient effects on electrical lines remain rather foggy or completely veiled.

Using the discrete parameter model reduces the state space of the system to a finite dimension, and the partial differential equations of the continuous (infinite-dimensional) time and space system into ordinary differential equations with a finite number of parameters. In this way the considered physical quantities, voltages and currents, become functions of time only. However, the final, closed-form (analytical) solution is not sufficiently accurate in the case of lumped parameter ladder circuits formed in reality by

Practical Methods for Analysis and Design of HV Installation Grounding Systems
ISBN 978-0-12-814460-2
https://doi.org/10.1016/B978-0-12-814460-2.00002-7

transmission line ground wires and the associated tower footing/grounding electrodes, especially in cases where the number of line spans is small.

During a ground fault in a power system, depending upon the fault location, at one of the line ends or somewhere along the line, one or two lumped parameter ladder circuits are formed through the ground wire(s) and tower grounding electrodes of the high-voltage (HV) overhead line. As a result, at one moment in the development of contemporary electric power systems, the exact mathematical model of these circuits has appeared as a practical need.

Extensive work has been undertaken, especially during the 1980s and 1990s, to model transmission lines for the purpose of ground fault current analysis. The advantages and drawbacks of a great number of the developed methods and procedures have been discussed in detail in Ref. [2]. Here we shall only point out that in their developments and improvements two tendencies can be clearly distinguished. On the one hand, an effort is made to make these methods more convenient for applications [3–8] due to the great number of cases that have to be analyzed. On the other hand, the authors try to improve the procedure further by including new, less important factors, but which, under some especially complex and unfavorable conditions, can contribute to obtaining an economically and technically acceptable solution [9,10].

Since the 1960s many authors have presented methods for solving the ladder circuit formed by a transmission line ground wire(s). The generally used approach was to represent the line ground wire by an equivalent distributed parameter ladder circuit and to solve it by using the well-known general line equations. However, the proposed methods give sufficiently accurate results only when a line, or its section(s), represented and solved in this manner, has a sufficiently large number of spans [2].

A second group of methods has been developed by using special matrix techniques that rely on modern computers. However, theoretical and practical considerations have shown that relative error propagation can be a problem when solving a large complex matrix. Thus further efforts for overcoming the problem have been directed toward the correct mathematical model, i.e., finding an analytical solution for lumped parameter ladder circuits.

In the 1980s, by using Kirchhoff's rules, the principle of superposition, and the summation of different geometric series, equations were derived that took into consideration the discrete nature of lumped parameter ladder circuits [3,10,11]. However, the procedure was not finalized, so that the obtained solutions were only partial, i.e., valid only for specific cases, from the standpoint of ladder circuit terminal conditions.

Finally, by deriving the general equations of the line represented by discrete parameters [12], the problem of determination of a general mathematical model for uniform discrete parameter ladder circuits had been definitively solved. Also the development of the analytical procedure that successively followed, one by one, all the phases of the actual physical process occurring before steady state was established. Thus the possibilities for obtaining a better insight into the physical essence of the resonant phenomena, and providing a better understanding and ease of research in of these in transmission lines, were made available, e.g., Refs. [13,14].

Subsequently, the mentioned general equations were used as a theoretical foundation in many research papers concerning mainly the problems of determination of critical ground fault position, e.g., Ref. [15], and ground fault current distribution along the lines that cannot be treated as homogeneous, e.g., Refs. [16,17]. Since the obtained calculation results were logical and within the expected limits, it seemed that the problem was definitively solved. The fact that the mentioned problems have not been the subject of further research in recent years certainly supports this assumption.

Unfortunately, the presented analytical procedure was not finalized and the derived general equations were not presented in the correct and final mathematical form. This certainly represents a defect from the standpoint of obtaining completely correct results by application of these equations. However, it has given a new foundation to an investigation of resonant phenomena and opened the possibility for a new investigation into transient states. The significance of this new approach lies in the fact that the development procedure strictly follows all the phases of the actual physical processes that precede establishing a steady state in a line. As such, it can be used not only for the analysis of the resonant phenomena [13,14], but also for the analysis and determination of transient overvoltages at any point along the line. Also, as such, a developed analytical procedure has undoubted value for educational purposes.

However, as a final achievement of the completely correct analytical procedure presented here, other, so far unknown, equations should be considered. By analogy to the general line equations they can be called the general equations of lumped parameter ladder circuits and applied to any electrical line or electrical device constructed as a uniform discrete parameter ladder network.

2.1.2 Lumped Parameter Circuit of an Electrical Line

It is well known that every electrical line for AC energy transmission or for telecommunications purposes can be represented by an arbitrary number of sections with lumped parameters. In this way, these lines, which in reality have distributed parameters, are fictitiously divided into an arbitrarily chosen number of equal sections with discrete parameters. The lumped parameter model of a line with the earth as return conductor and with a total of N fictitious sections of discretization, from now on simply called "line sections," is presented in Fig. 2.1.

The notation has the following meaning:

- U_0—voltage representing external electromotive forces,
- U_N—voltage appearing at the output of the line,
- I_0, I_1, ..., I_N—currents of individual line sections,
- Z_N—load impedance, and
- N—adopted total number of line sections.

Here it is necessary to mention that the quantities that represent phasors are written implying $e^{j\omega t}$ time dependence throughout.

As shown, the lumped parameter model of a homogeneous electrical line is composed of a chain of identical π-networks each containing one impedance Z_S and two admittances $Y_S/2$, which are located at the ends of each line section obtained by discretization. The impedance Z_S and admittance Y_S are in the general case determined by:

$$Z_S = (r + j\omega l)L/N, \tag{2.1}$$

$$Y_S = (g + j\omega c)L/N, \tag{2.2}$$

where parameters r, 1, g, and c represent per unit length series resistance, series inductance, shunt conductance, and shunt capacitance of the line, respectively, while ω and L are the angular frequency of the sinusoidal voltage U_0 and total line length, respectively.

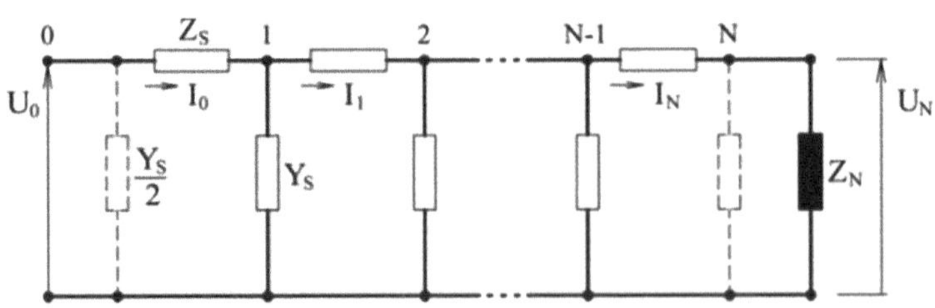

Figure 2.1 A line represented by discrete parameters [12].

The parameters r, 1, g, and c are dependent on the geometry and insulation of the conductor, the electric, dielectric, and magnetic properties of the materials used in constructing the line. The geometrical factors are independent of frequency, but the physical properties (resistivity, dielectric constant, and permeability) are in general functions of frequency. Thus, for simplicity, we will consider here the electrical quantities only at a fixed frequency.

For the purpose of developing the analytical procedure, let us assume that the arbitrarily chosen line section length is so small that the admittance $Y_S/2$ at the line ends (denoted by dashed lines) can be neglected. At the same time, this means that I_0 represents the sending/input current, while I_N represents the receiving/output current of the presented lumped parameter ladder circuit (Fig. 2.1). The lumped parameter circuit obtained by using this approximation has a form that exactly corresponds to the existing circuit formed through a transmission line ground wire(s) and its connections with tower footing electrodes.

Also, according to Fig. 2.1, voltage U_0, fed by the generator having negligible internal impedance, is applied to the input end of the presented line model, while the output end is terminated by impedance Z_N.

2.1.3 Model of an Infinitely Long Line

2.1.3.1 Input Impedance of an Infinite Line Model

We will start with a case that is far from reality but that enables a better insight into the interplay of the relevant parameters to be obtained. Thus initially we will assume that the considered line has been prolonged so that its end is somewhere at infinity. In that case, we have the circuit in Fig. 2.2, where point n corresponds to point N of Fig. 2.1.

Since it is assumed that the line is infinite, input impedance of this line seen from node 1 has the same value as at the beginning of line 0. If we

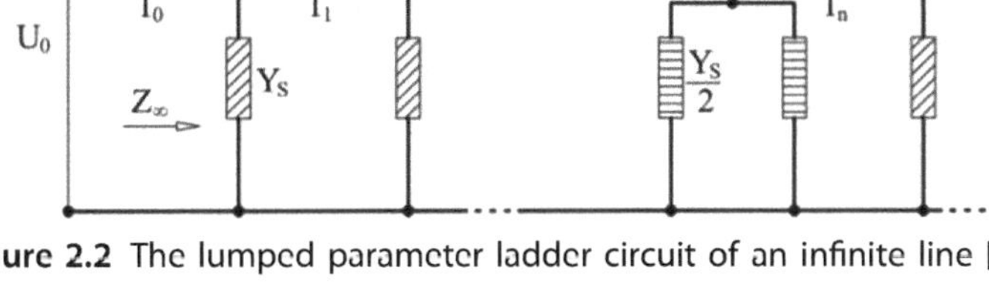

Figure 2.2 The lumped parameter ladder circuit of an infinite line [12].

denote this impedance by Z_∞, we can write the following equations for the currents and voltages of the first and second line sections:

$$\begin{aligned} Z_\infty I_0 &= Z_S I_0 + Z_\infty I_1 \\ I_0 &= I_1 + Y_S Z_\infty I_1. \end{aligned} \tag{2.3}$$

By eliminating I_0 and I_1 from these equations, we obtain a quadratic equation. The positive root of this equation gives the analytical expression for the input impedance of the lumped parameter ladder circuit of an infinite line. It is determined by:

$$Z_\infty = \frac{Z_S}{2} + \sqrt{\frac{Z_S}{Y_S} + \frac{Z_S^2}{4}}. \tag{2.4}$$

According to the given expression, the input impedance of an infinite line model and the line characteristic impedance are not identical. However, in the case of a transmission line and power frequency, we can use the following approximation:

$$Z_\infty \approx \frac{Z_S}{2} + \sqrt{\frac{Z_S}{Y_S}} = \frac{Z_S}{2} + Z_C, \tag{2.5}$$

where Z_C is the line characteristic impedance.

2.1.3.2 Currents Along the Lumped Parameter Model of an Infinite Line

In this case the input impedance seen from any node along the line model 1, 2, ..., n, ... (Fig. 2.2) toward its end has a constant value equal to impedance Z_∞. This means that the ratio of the currents of any two neighboring sections has the same value at all nodes of the presented circuit. Thus, according to Kirchhoff's rules, the following relation can be written:

$$k = \frac{I_n}{I_{n+1}} = 1 + Y_S Z_\infty. \tag{2.6}$$

Then, the relations between currents of individual sections of an infinite line model can be represented by the terms of an infinite geometric series with the argument equal to k, or:

$$\begin{aligned} I_{N-1} &= kI_N, \\ I_{N-2} &= kI_{N-1} = k^2 I_N, \\ I_{N-3} &= \cdots = k^3 I_N, \\ &\vdots \\ I_0(\infty) &= \cdots = k^{N-1} I_N, \end{aligned} \tag{2.7}$$

where $I_0(\infty)$ is the current of the first section, which is obviously determined by:

$$I_0(\infty) = \frac{U_0}{Z_\infty}. \tag{2.8}$$

If we multiply each of Eq. (2.7) by impedance Z_∞, we will see that the potentials of individual nodes along an infinite line model are also mutually related as the terms of a geometrical series. This certainly means that if U_0 is known we can easily determine any current and voltage along the circuit in Fig. 2.2.

2.1.3.3 Transfer Impedance of the Lumped Parameter Ladder Circuit

For lumped parameter ladder circuits it is possible to define one more impedance that also represents an important characteristic for this type of circuit. This impedance enables us to represent the equivalent circuit of Fig. 2.1, from the standpoint of the current and voltage appearing at its end by a very simple equivalent circuit shown in Fig. 2.3.

The analytical expression for this impedance, denoted by Z_{0N}, can be determined on the basis of Ohm's law and the circuits of Figs. 2.1 and 2.3. At first, according to the circuit of Fig. 2.1, we can write:

$$U_0 = \left(\frac{I_0}{I_N} + \frac{I_1}{I_N} + \cdots \frac{I_{N-1}}{I_N} + 1\right) Z_S I_N + Z_N I_N, \tag{2.9}$$

while, according to the circuit of Fig. 2.3, we have:

$$U_0 = Z_{0N} I_N + Z_N I_N. \tag{2.10}$$

On the basis of Eqs. (2.9) and (2.10) and under the condition that currents and voltages at the line ends remain unchanged, the transfer impedance of the line model in general is determined by:

$$Z_{0N} = \frac{\sum_{0}^{N-1} I_n}{I_N} Z_S. \tag{2.11}$$

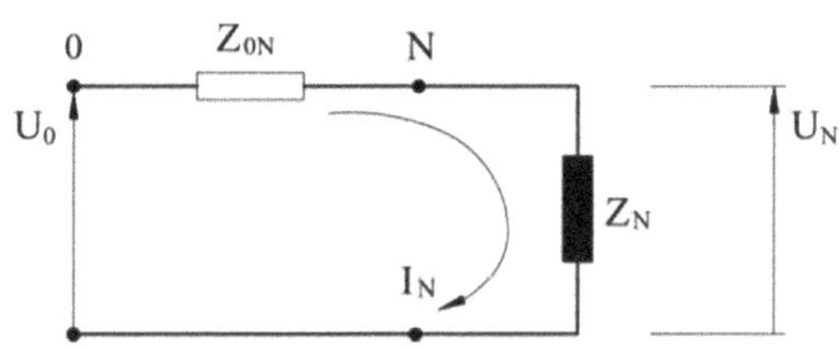

Figure 2.3 Transfer impedance of the lumped parameter ladder circuit [12].

As we can see, for determining the transfer impedance it is necessary to know the currents of all individual sections of the line model.

If we now assume that the impedance connected at the end of the line model (Fig. 2.1) is equal to impedance $Z_\infty - Z_S$ ($Z_N = Z_\infty - Z_S$), it is evident on the basis of the circuit shown in Fig. 2.1 that the distribution of currents and voltages along the line model will be as in the case of infinite line model. Therefore, in accordance with Eqs. (2.7) and (2.11), it follows that the transfer impedance of the line model closed by $Z_\infty - Z_S$ is given by:

$$Z_{0N}(Z_N = Z_\infty - Z_S) = \frac{k^N - 1}{k - 1} Z_S. \tag{2.12}$$

However, in general the impedance at the end of the line, Z_N, can take any value between zero and infinity ($0 \leq Z_N \leq Z_\infty$). Thus our consideration should be continued to involve these cases. With the aim of simplifying the problem, we will reduce the number of involved quantities by assuming at the beginning that this impedance is equal to zero, $Z_N = 0$.

2.1.4 Short Circuit at the End of the Line Model

Keeping in mind that we are dealing with a passive electrical circuit (with constant parameters), for the purpose of finally solving the considered problem we will use the principle of superposition, as follows. At first we shall assume that voltage U_0 is applied to a line that we have fictitiously extended, without changing its parameters, to make it infinitely long. Then, for obtaining a short-circuited line or zero voltage at the end of the line ($U_N = 0$), it is necessary to impose a voltage U_{NS1} at the end of the line that is equal in magnitude but of opposite sign to the one that would appear at this point under the assumption that the line is infinite. Now, let us suppose that this imposed voltage is the only one applied to the line and that the line is infinite, but that now we look at it from the line end toward the beginning, which is somewhere in infinity. If we assume that voltage U_0 connected to the line remains constant, e.g., that it is generated by an ideal voltage source (having negligible internal impedance), then the voltage produced by U_{NS1} at the beginning of the line, U_{01}, must be canceled by superposing U_{0S1} of the same magnitude but of the opposite sign. Now, let us consider the current and voltage distribution along the line due to U_{0S1} only, also assuming that the line is infinite, and repeat the whole procedure as before with U_0 and U_{NS1}. The cancellation voltages that have to be superimposed gradually become smaller and smaller, but to satisfy the voltage conditions at the terminal points of the line, they have be taken into consideration until they become negligibly small. At the moment when the need for any new superimposed voltage disappears, the voltages and currents along the line model become periodical (sinusoidal) functions of the frequency equal to the supply frequency, and we can say that the steady state on the line model is established. For an actual (with distributed parameters) energized power line this is the state that appears when traveling waves completely disappear or when voltages and currents at any point along the line become mutually related by Ohm's law.

The previously described procedure is illustrated graphically in Fig. 2.4, under the assumption that the considered line model has six equal sections and that Z_S (Eq. 2.1) and Y_S (Eq. 2.2) have only real parts. If we were to take into account the inductance and capacitance of the line, all voltages

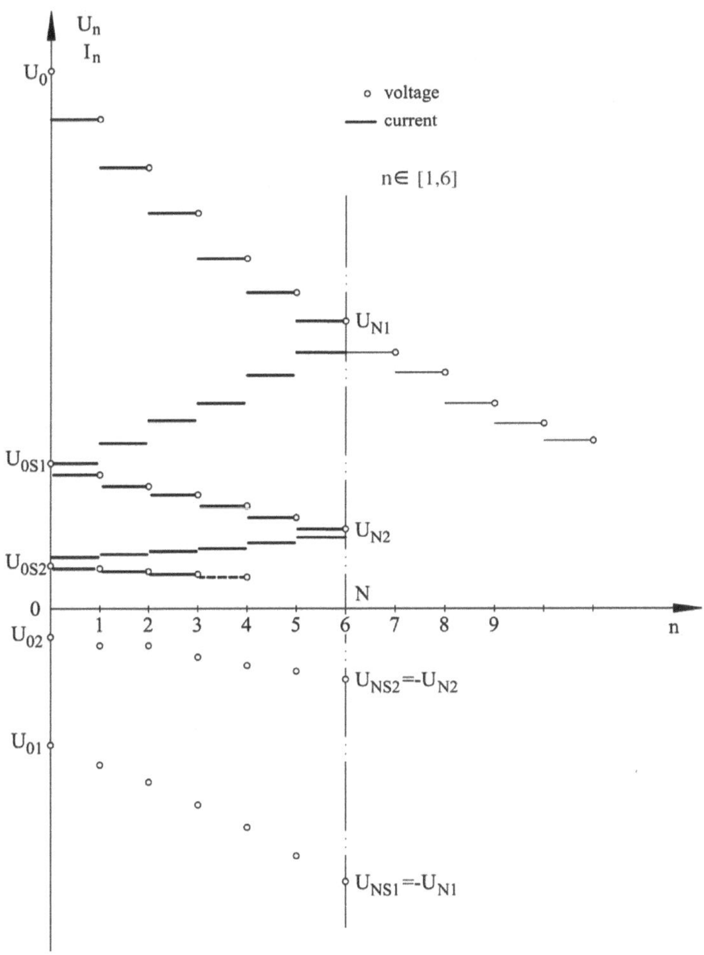

Figure 2.4 Graphical representation of the described analytical procedure [12].

would be shifted by phase angles and could be represented only by phases in a three-dimensional space.

According to Fig. 2.4, the voltages that have to be superimposed at the beginning of the line would appear at the beginning of the 12th, 24th, 36th, ... sections, while the voltages that should be superimposed at the end of the modeled line would appear at the beginning of the 6th, 18th, ... sections, if the line were infinite. Also from this diagram it follows that irrespective of the theoretically infinite number of voltages that have to be superimposed, the sum of the voltages at the beginning of the line model remains equal to U_0, while the sum of the voltages at its end remains equal to zero. Due to the assumptions made, it is not difficult to note that these voltages on a line model create voltages and currents that correspond to the traveling waves along actual transmission lines. This can be proved on the basis of the following consideration.

Any (mth) of the voltages applied to the beginning of the line (Fig. 2.4), U_{0Sm}, create at the opposite line end the following voltage:

$$U_N(U_{0Sm}) = U_{0Sm}k^{-N}. \tag{2.13}$$

When parameter k in Eq. (2.13) is substituted by expression (2.6), where Y_S is expressed by using the number of line sections, N, and the corresponding line length, L, instead of Eq. (2.13), we obtain:

$$U_N(U_{0Sm}) = U_{0Sm}\left(1+\frac{L\gamma Z_\infty}{N}\right)^{-\frac{N}{2}}. \tag{2.14}$$

If we now assume that the number of line sections, N, for an unchanged line length increases, impedance Z_∞, determined by Eq. (2.4), becomes equal to the characteristic line impedance, or:

$$\lim_{N\to\infty}\left(\frac{zL}{2N}+\sqrt{\frac{z}{y}+\frac{z^2L^2}{4N^2}}\right)=\sqrt{\frac{z}{y}}. \tag{2.15}$$

From Eq. (2.14) it becomes clear that the voltage at the end of the line is:

$$\lim_{N\to\infty} U_N(U_{0Sm}) = U_{0Sm}e^{-\gamma L}, \tag{2.16}$$

where γ represents the line propagation constant (e.g., Ref. [1]) determined by:

$$\gamma = \sqrt{zy} = \sqrt{(r+j\omega l)(g+j\omega c)}. \tag{2.17}$$

In a similar manner it is possible to show that the voltages applied to the end of the line create reflected traveling waves on an actual (having distributed parameters) line.

As is shown, each superimposed voltage "sees" the same input impedance as it propagates. Also each voltage wave is associated with the corresponding current wave. The ratio of these two waves at any point along the line model is equal to impedance Z_∞.

The stationary currents and voltages of the individual sections, i.e., at individual nodes along the line model (Fig. 2.1), can be determined on the basis of the principle of superposition, by summing all currents/voltages, which are produced at these places by the voltage U_0 and each of the superimposed voltages. As voltage U_0 and each of the superimposed voltages create currents and potentials distributed along the line model as if the line were infinite, determination of the stationary currents of individual line sections can be achieved by forming and summing the corresponding geometric series, as follows.

Currents resulting from voltage U_0 and from each of the superimposed voltages are distributed in such a manner that all the currents that result from certain voltages form one row, from the first to the last section of the line model. The rows formed in this manner are arranged sequentially one below the other, starting at the one created by voltage U_0 and followed by the sequence of rows produced by the sequence of the superimposed voltages (Fig. 2.4). In this way we create for each line section a column of currents that belong to it. If we divide each of these currents by the current produced by voltage U_0 at the beginning of the line, under the assumption that the line is infinitely long, $I_0(\infty)$, we obtain:

$$\begin{array}{ccccccc}
1 & k^{-1} & \cdots & k^{-n+1} & \cdots & k^{-N+2} & k^{-N+1} \\
k^{-2N+1} & k^{-2N+2} & \cdots & k^{-2N+n} & \cdots & k^{-N-1} & k^{-N} \\
k^{-2N} & k^{-2N-1} & \cdots & k^{-2N-n+1} & \cdots & k^{-3N+2} & k^{-3N+1} \\
k^{-4N+1} & k^{-4N+2} & \cdots & k^{-4N+n} & \cdots & k^{-3N-1} & k^{-3N} \\
k^{-4N} & k^{-4N-1} & \cdots & k^{-4N-n+1} & \cdots & k^{-5N+2} & k^{-5N+1} \\
- & - & \cdots & - & \cdots & - & - \\
- & - & \cdots & - & \cdots & - & - \\
- & - & \cdots & - & \cdots & - & -
\end{array} \tag{2.18}$$

The fact that all superposed voltages, both at the beginning and at the end of the line, produce voltages along the line, whose value decreases from the beginning to the line end (Fig. 2.4), means that all currents along the line flow in the same direction, i.e., toward the line end. Therefore we can determine the value of the stationary current in certain sections if we sum up all the currents in the corresponding column. Since each of the columns contains two infinite geometrical series (Eq. 2.18) and since the argument of these series is less than one ($|k^{-2N}| < 1$ for each $N \geq 1$, because $|k| > 1$), these series are convergent and their sums can be represented in closed form. According to this, for the first column we obtain:

$$\frac{k}{k^{2N}} \sum_{n=0}^{\infty} \left(\frac{k}{k^{2N}} \right)^n = \frac{k}{k^{2N} - 1}, \tag{2.19}$$

$$\frac{1}{k^{2N}} \sum_{n=0}^{\infty} \left(\frac{1}{k^{2N}} \right)^n = \frac{1}{k^{2N} - 1}. \tag{2.20}$$

If we do the same with the series of each of the previously formed columns, and then show each of them as a row, starting from the first, we obtain Eq. (2.21).

As can be seen, each row has three terms in the sum, which, together with the corresponding terms of the remaining rows, form three finite geometrical series whose terms are arranged in the columns (Eq. 2.21).

$$\begin{aligned}
S_0 &= 1 + \frac{k}{k^{2N} - 1} + \frac{1}{k^{2N} - 1}, \\
S_1 &= k^{-1} + \frac{k^2}{k^{2N} - 1} + \frac{k^{-1}}{k^{2N} - 1}, \\
S_2 &= k^{-2} + \frac{k^3}{k^{2N} - 1} + \frac{k^{-2}}{k^{2N} - 1}, \\
&\vdots \\
S_n &= k^{-n} + \frac{k^{n+1}}{k^{2N} - 1} + \frac{k^{-n}}{k^{2N} - 1}, \\
&\vdots \\
S_{N-1} &= k^{-N+1} + \frac{k^N}{k^{2N} - 1} + \frac{k^{-N+1}}{k^{2N} - 1}.
\end{aligned} \tag{2.21}$$

Therefore the sum of the terms of all previously formed columns (Eq. 2.18) can be written in the following compact form:

$$\sum_{n=0}^{N-1} S_n = \frac{k}{1-k}. \tag{2.22}$$

As can be seen, Eq. (2.22) represents the sum of an infinite geometrical series with k as argument. It means that the sum of all currents along any line model short circuited at its end for certain U_0 depends only on parameter k (Eq. 2.6), i.e., on line parameters r, l, g, and c. For the same line parameters this sum is constant and, according to Eq. (2.22), equal to the sum of all currents appearing along the line model under the assumption that it is infinitely prolonged. Thus this sum is not dependent on the adopted number of line model sections, N. In the case of an actual line of any length and short circuited at its end, this means that current integral along this line is equal to the current integral along the same line but prolonged to infinity, and it is not dependent on the line length.

By taking into consideration the way in which the initial columns (Eq. 2.18) were formed, it is clear that the products of the sum of terms of any of the columns and the current $I_0(\infty)$ give the stationary current in the corresponding line section.

If we order in an analogous manner the voltages of individual nodes (0, 1, 2, N) along the line model and divide them by U_0, we obtain:

$$\begin{array}{cccccc} 1 & k^{-1} & \cdots & k^{-n+1} & \cdots & k^{-N} \\ -k^{-2N} & -k^{-2N+1} & \cdots & -k^{-2N+n-1} & \cdots & -k^{-N} \\ k^{-2N} & k^{-2N-1} & \cdots & k^{-2N-n+1} & \cdots & k^{-3N} \\ -k^{-4N} & -k^{4N+1} & \cdots & -k^{-4N+n-1} & \cdots & -k^{3N} \\ - & - & \cdots & - & \cdots & - \\ - & - & \cdots & - & \cdots & - \\ - & - & \cdots & - & \cdots & - \end{array} \tag{2.23}$$

By summing up the columns in Eq. (2.23), where each contains two infinite geometrical series, we obtain:

$$\begin{aligned} S'_0 &= 1, \\ S'_1 &= k^{-1} - \frac{k - k^{-1}}{k^{2N} - 1}, \\ S'_2 &= k^{-2} - \frac{k^2 - k^{-2}}{k^{2N} - 1}, \\ &\vdots \\ S'_n &= k^{-n+1} - \frac{k^n - k^{-n}}{k^{2N} - 1}, \\ &\vdots \\ S'_{N-1} &= k^{-N+1} - \frac{k^N - k^{-N}}{k^{2N} - 1}, \\ S'_N &= 0. \end{aligned} \tag{2.24}$$

On the basis of the presented procedure it is clear that the product of certain of these sums and voltage U_0 is equal to the stationary voltage at the corresponding node along the line model.

According to Eq. (2.21), the analytical expression for the current of an arbitrary, nth, section is:

$$I_n = k^{-n}\left(1 + \frac{1 + k^{2n+1}}{k^{2N} - 1}\right) I_0(\infty). \tag{2.25}$$

Since the current of the first section is also determined, I_0 (2.21), the input impedance is:

$$Z_i(Z_N = 0) = \frac{k^{2N} - 1}{k^{2N} + k} Z_\infty. \tag{2.26}$$

Change (decrease) in the input impedance in comparison with the case when the line model is closed at the end by impedance $Z_\infty - Z_S$ is:

$$\Delta Z_i = Z_i(Z_N = Z_\infty - Z_S) - Z_i(Z_N = 0) = -\frac{1 + k}{k^{2N} + k} Z_\infty. \tag{2.27}$$

Finally, when expressions (2.21) and (2.22) are determined, then, from Eq. (2.11), we obtain that the transfer impedance is:

$$Z_{0N}(Z_N = 0) = \frac{k^{2N+1} - k}{k^N - k^{N+2}} Z_S. \tag{2.28}$$

By summing the terms of the two infinite geometrical series (Eq. 2.23) it is possible to determine the stationary voltage at any node along the considered line model. The voltage at an arbitrary node, n, of the line model is determined by:

$$U_n = k^{-n}\left(1 - \frac{k^{2n} - 1}{k^{2N} - 1}\right) U_0. \tag{2.29}$$

As can be seen the derived analytical expressions can be used for any type of AC line independently of the value of ω and line parameters r, l, g, and c. Also, as can be seen each voltage wave "sees" an impedance as it propagates.

2.1.5 Arbitrarily Loaded Line Model

Let us now consider the line model under an arbitrary load, or when the load impedance Z_N has any value from zero to infinity ($0 < Z_N < \infty$). As in the previous case, for such an operating regime we also must appreciate the end conditions. Thus voltage at the end of the line model, U_N, in this case should have certain value equal to $Z_N I_N$. To fulfill this condition it is necessary to change the value of the first voltage, which we superimpose at the end of the line. We shall accomplish this by multiplying U_{NS} (Fig. 2.4) by a coefficient that we will denote by α. This coefficient at present is an unknown quantity, but will be determined later. Since we are dealing with a circuit having constant parameters, this simultaneously means that, following the previously described procedure, we must also multiply all the other superimposed voltages by the same coefficient. Then, the relations between the currents of individual sections along the line model can be presented in a similar manner to the previous case, or:

$$\begin{array}{ccccccc}
1 & k^{-1} & \cdots & k^{-n+1} & \cdots & k^{-N+2} & k^{-N+1} \\
\alpha k^{-2N+1} & \alpha k^{-2N+2} & \cdots & \alpha k^{-2N+n} & \cdots & \alpha k^{-N-1} & \alpha k^{-N} \\
\alpha k^{2N} & \alpha k^{-2N-1} & \cdots & \alpha k^{-2N-n+1} & \cdots & \alpha k^{-3N+2} & \alpha k^{-3N+1} \\
\alpha k^{-4N+1} & \alpha k^{4N+2} & \cdots & \alpha k^{-4N+n} & \cdots & \alpha k^{-3N-1} & \alpha k^{-3N} \\
\alpha k^{-4N} & \alpha k^{-4N-1} & \cdots & \alpha k^{-4N-n+1} & \cdots & \alpha k^{5N+2} & \alpha k^{-5N+1} \\
- & - & \cdots & - & \cdots & - & - \\
- & - & \cdots & - & \cdots & - & - \\
- & - & \cdots & - & \cdots & - & -
\end{array} \tag{2.30}$$

The sums of these columns, obtained by using the expression for the sum of infinite geometrical series, are:

$$\begin{aligned}
S_0 &= 1 + \alpha \frac{k+1}{k^{2N}-1}, \\
S_1 &= k^{-1} + \alpha \frac{k^2 + k^{-1}}{k^{2N}-1}, \\
S_2 &= k^{-2} + \alpha \frac{k^3 + k^{-2}}{k^{2N}-1}, \\
&\vdots \\
S_n &= k^{-n} + \alpha \frac{k^{n+1} + k^{-n}}{k^{2N}-1}, \\
&\vdots \\
S_{N-1} &= k^{-N+1} + \alpha \frac{k^N + k^{-N+1}}{k^{2N}-1}.
\end{aligned} \tag{2.31}$$

By analogy with the procedure described in the previous case, for the relations between the voltages at individual nodes along the line model we now have the following columns:

$$\begin{array}{ccccccc}
1 & k^{-1} & \cdots & k^{-n+1} & \cdots & k^{-N} \\
\alpha k^{-2N} & \alpha k^{-2N+1} & \cdots & \alpha k^{-2N+n-1} & \cdots & \alpha k^{-N} \\
\alpha k^{-2N} & \alpha k^{-2N-1} & \cdots & \alpha k^{-2N-n+1} & \cdots & \alpha k^{-3N} \\
\alpha k^{-4N} & \alpha k^{4N+1} & \cdots & \alpha k^{-4N+n-1} & \cdots & \alpha k^{-3N} \\
- & - & \cdots & - & \cdots & - \\
- & - & \cdots & - & \cdots & - \\
- & - & \cdots & - & \cdots & -
\end{array} \tag{2.32}$$

By summing up the terms of each of the presented columns we obtain:

$$\begin{aligned}
S'_0 &= 1, \\
S'_1 &= k^{-1} + \alpha\frac{k + k^{-1}}{k^{2N} - 1}, \\
S'_2 &= k^{-2} + \alpha\frac{k^2 + k^{-2}}{k^{2N} - 1}, \\
&\vdots \\
S'_n &= k^{-n} + \alpha\frac{k^n + k^{-n}}{k^{2N} - 1}, \\
&\vdots \\
S'_N &= (1 + \alpha)k^{-N}.
\end{aligned} \tag{2.33}$$

By summing up the currents produced in certain line sections by voltage U_0 and all the superimposed voltages, according to the principle of superposition, we obtain the stationary current of that section. According to this, the total currents of the first and last line section are:

$$I_0(0 \leq Z_N \leq \infty) = \left(1 - \alpha\frac{1 + k}{1 - k^{2N}}\right)\frac{U_0}{Z_\infty}, \tag{2.34}$$

$$I_N(0 \leq Z_N \leq \infty) = k^{-N+1}\left(1 - \alpha\frac{1 + k^{2N-1}}{1 - k^{2N}}\right)\frac{U_0}{Z_\infty}. \tag{2.35}$$

The relationship between the voltages at the beginning and end of the line, significant for determining the conditions for power transfer under constant voltage, according to Eq. (2.33), is determined by:

$$U_N = (1 + \alpha)k^{-N} U_0. \tag{2.36}$$

Then, on the basis of Fig. 2.1, the current and voltage at the line end are related through:

$$U_N = Z_N I_N. \tag{2.37}$$

When the expressions for I_N (2.35) and U_N (2.36) are introduced in Eq. (2.37), it follows that coefficient α is in general determined by:

$$\alpha = \frac{(k^{2N} - 1)(Z_\infty - kZ_N)}{(k^{2N} - 1)Z_\infty + (k^{2N} + k)Z_N}. \tag{2.38}$$

Finally, by introducing the corresponding values for the load impedance, Z_N, into expression (2.38), expressions (2.34–2.36) can be modified to cover any operating conditions, including a line model under no load ($Z_N = \infty$).

2.1.6 General Equations of the Lumped Parameter Ladder Circuits

The well-known general line equations describe the relationship between voltages and currents at the beginning and end of a line. Here, by using expressions (2.34), (2.35), and (2.37), after certain algebraic manipulations, we obtain the following system of equations:

$$\begin{aligned} &1.\ U_0 = \frac{k^N + k^{-N+1}}{1+k} U_N + \frac{(k^N - k^{-N})Z_\infty}{1+k} I_N, \\ &2.\ I_0 = \frac{k^N - k^{-N+2}}{(1+k)Z_\infty} U_N + \frac{k^N + k^{-N+1}}{1+k} I_N. \end{aligned} \tag{2.39}$$

This system of equations gives the relationship between voltages and currents at the ends of a lumped parameter ladder circuit of an arbitrary size ($1 < N < \infty$) and under any terminal conditions ($0 \leq Z_N \leq \infty$). Thus it can be stated that the derived equations represent the general equations of discrete parameter ladder circuits.

On the basis of expressions (2.15–2.17) it is not difficult to show that by an infinite increase in the number N ($N \to \infty$) the derived Eq. (2.39) are approaching the mathematical form, known as the general line equations.

This is certainly a new way of obtaining these well-known equations and also the analytical proof of the accuracy of the presented analytical procedure. From the standpoint of mathematics it can be said that the problem in the past was solvable only by solving partial differential equations, but is solved now by forming and summing specific infinite and finite geometric series and by using limits returning to the continuous functions.

Also, on the basis of Eq. (2.39), it is possible to obtain solutions for different practical problems associated with the electrical devices that have discretized parameters such as transmission line ground wires, transformer windings, electrical filters, insulator chains, etc. In the case of a circuit formed by transmission line ground wire(s), N represents the number of line spans.

By using the first of Eq. (2.39) the transfer impedance of lumped parameter ladder circuits is in general determined by:

$$Z_{0N} = \frac{U_0}{I_N} = \frac{(k^N + k^{-N+1})Z_N + (k^N - k^{-N})Z_\infty}{1+k} \tag{2.40}$$

On the basis of Eq. (2.40), any lumped parameter ladder circuit (Fig. 2.1) can be, from the standpoint of U_0, U_N, and I_N, substituted by the very simple circuit presented in Fig. 2.3. This certainly reduces the necessary calculations for solving any kind of practical problems in which U_N or I_N appear as unknown quantities.

Also, by applying Eq. (2.39) to an actual transmission line, the accuracy of these equations is increased if the freely chosen section number, N, is increased. Thus, by an unlimited increase in the number of discrete parameters, the accuracy of these equations becomes practically equal to the accuracy of the well-known general line equations. Thus Eq. (2.39) can also be used as an alternative to the significantly older and well-known general line equations. However, it is necessary to mention that for the same value of U_0, by disregarding $Y_S/2$ at the ends of the line model (Fig. 2.1), the values obtained for I_0, U_N, and I_N, by using Eq. (2.39), are slightly higher.

Then, it is necessary to mention again that the general line equations cannot be an alternative to Eq. (2.39) in cases where the length of a transmission line or length of one its sections is only several (a few) spans. As such, Eq. (2.39) can be used for solving all theoretical and practical problems dealing with the steady states of transmission lines and their models with lumped parameters. Also these equations can be used for calculations necessary for determining elements of electrical filters constructed in the form of an uniform lumped parameter ladder network.

The developed analytical procedure and resulting general Eq. (2.39) are obtained by using a lumped parameter ladder network formed in reality by HV transmission line ground wire(s). These equations are necessary for correct determination of ground fault current distribution at any tower along a transmission line. Also they can analyze electrical quantities along any line in steady state by applying relatively simple mathematical operations and with a desired degree of accuracy.

Finally, the presented analytical procedure is very convenient for analysis of the resonant phenomena in transmission lines and their discrete parameter models [13,14,16]. This convenience originates from the fact that in a theoretically infinite number of traveling waves, which participate in establishing a steady state in a line where the resonant conditions are to be analyzed, it is sufficient to observe the phase angle of only one voltage wave at the beginning of the line and the phase angle of the same wave, but after it has traveled to the end of the line and returned as reflected to the beginning of the line. When the phase angles of the phasors representing a direct and corresponding reflected voltage at the line beginning are such that they are summed algebraically, resonance oscillations appear. By using this observation as a theoretical foundation, relatively simple analytical expressions are derived for the resonant frequencies in the case of a long radial transmission line with static VAR compensation [12]. The relatively simple mathematical form of these expressions enables a more direct insight into the interplay of the relevant parameters and allows a more correct physical interpretation of the resonant phenomena than it was before. As is well known, they appear mainly as a consequence of the existence of higher harmonics in modern electric power networks [13].

2.1.7 General Equations of the Line Represented by Discrete Parameters

When we consider the entire circuit presented in Fig. 2.1, i.e., when we do not disregard the admittances $Y_S/2$ at its ends, we have a chain of identical π-cells. Each π-cell is composed of one impedance Z_S and two admittances $Y_S/2$, which in nodes 1, 2, 3, …, $N-1$ form admittances Y_S. Voltages and currents at the ends of the circuit formed in such a manner, according to the theory of periodical networks, are related by the following equations:

$$\begin{aligned} &1.\ U_0 = ch\gamma_C U_N + Z_C sh\gamma_C I_N \\ &2.\ I_0 = \frac{sh\gamma_C}{Z_C} U_N + ch\gamma_C I_N, \end{aligned} \tag{2.41}$$

where Z_C is the characteristic impedance of the line circuit and γ_C is the characteristic transfer constant of the line circuit.

The characteristic impedance Z_C is defined as the impedance that, when connected at the line end, results in the same impedance appearing at the input terminals. On the basis of the already defined Z_∞ (Eq. 2.4) and k (Eq. 2.6) and according to Figs. 2.1 and 2.3, it can be presented by:

$$Z_C = \frac{2Z_\infty}{Y_S Z_\infty + 2} = \frac{2Z_\infty}{1+k} \tag{2.42}$$

The characteristic transfer constant, γ_C, is the natural logarithm of the ratio of voltages at the input and output (or the currents at the input and output) of the circuit closed by its characteristic impedance. When we close the circuit presented in Fig. 2.1 by its characteristic impedance, it is easy to show that it is:

$$\frac{U_0(Z_N = Z_C)}{U_N(Z_N = Z_C)} = \frac{I_0(Z_N = Z_C)}{I_N(Z_N = Z_C)} = k^N \tag{2.43}$$

On the basis of the given definition for γ_C and Eq. (2.43) we obtain:

$$\gamma_C = \ln k^N = N \ln k \tag{2.44}$$

Finally, when we substitute γ_C in Eq. (2.41) with Eq. (2.44) we obtain the following system of equations:

$$\begin{aligned} &1.\ U_0 = \frac{k^N + k^{-N}}{2} U_N + \frac{(k^N - k^{-N}) Z_C}{2} I_N \\ &2.\ I_0 = \frac{k^N - k^{-N}}{2Z_C} U_N + \frac{k^N + k^{-N}}{2} I_N \end{aligned} \tag{2.45}$$

Eq. (2.45) correspond to the circuit involving all the elements obtained by the discretization of a homogeneous line, as shown in Fig. 2.1. Thus they can be called the general equations of the line presented by discrete parameters. As such, they are very convenient for analyzing the grounding effects of medium-voltage (MV) cable lines with uncovered metal sheaths, as will be shown later.

2.1.8 Transient Voltages

Among the applications of analyses of electromagnetic transients in power systems the principal ones are those intended for evaluation of conditions wherein either system voltages are forced to exceed the designed nominal levels or circuit breakers are stressed during switching operations.

For considering these problems, instead of the geometrical series presented by Eq. (2.23), we have the following:

$$\begin{array}{ccccccccc}
1 & k^{-1} & \cdots & k^{-n+1} & \cdots & k^{-N} \\
\beta k^{-2N} & \beta k^{-2N+1} & \cdots & \beta k^{-2N+n-1} & \cdots & \beta k^{-N} \\
-\beta k^{-2N} & -\beta k^{-2N-1} & \cdots & -\beta k^{-2N-n+1} & \cdots & -\beta k^{-3N} \\
\beta^2 k^{-4N} & \beta^2 k^{-N+1} & \cdots & \beta^2 k^{-4N+n-1} & \cdots & \beta^2 k^{-3N} \\
-\beta^2 k^{-4N} & -\beta^2 k^{-4N-1} & \cdots & -\beta^2 k^{-4N-n+1} & \cdots & -\beta^2 k^{-5N}, \\
- & - & \cdots & - & \cdots & - \\
- & - & \cdots & - & \cdots & - \\
- & - & \cdots & - & \cdots & -
\end{array} \tag{2.46}$$

where β is the reflection coefficient at the end of the line model.

The reflection coefficient at the line model end is, according to Ref. [13], determined by:

$$\beta = \frac{kZ_N - Z_\infty}{Z_N + Z_\infty} \tag{2.47}$$

In cases where internal impedance of the voltage source U_0, Z_0, is not negligible, the reflection coefficient at the line beginning β_0 is determined by the same expression, but instead of Z_N one should write Z_0.

The developed procedure and expressions (2.45) and (2.47) can be used when we wish to determine the voltage at any node (1, 2, 3, N) along the line model during a transient state. To obtain the maximal/critical transient voltage at a certain node it is sufficient to sum, one by one, the members of the corresponding column, starting from the one in the first row under the assumption that its value at that moment is equal to its amplitude. For example, to obtain the maximal possible transient voltage at the end of a single-phase and no-loaded line we have, under the assumption of an infinite bus source, the following relation:

$$U_N = 2U_N k^{-N}(1 - k^{2N} + k^{4N} - k^{6N} + \cdots). \tag{2.48}$$

The amplitude of the first traveling wave (row) can appear at the line end at any moment during its reflection from the line end. Thus by transforming Eq. (2.48) in the time domain, the maximum overvoltage is determined by:

$$U_{Nm}(t) = 2U_a k^{-N} \cos \omega(t - t_a);\ 0 \leq t_a \leq T, \tag{2.49}$$

where U_a is the amplitude of the phase voltage at the line beginning, U_{Nm} is the maximal transient overvoltage at the line end, T is the time of traveling wave propagation through the line, t is the time counted from the appearance of the first traveling wave at the line end, and t_a is the time counted from the appearance of the voltage amplitude at the line end.

In other words, the maximal possible overvoltage occurs when the first traveling wave amplitude appears at the line beginning with a delay not smaller than T and not larger than $2T$.

Here it should be mentioned that the skin effect can be taken into account through the corresponding value of the phase conductor resistance per unit length, r. By introducing the corresponding line parameters, a similar calculation procedure can be used for transient analysis in DC transmission lines. Then, the developed method can be relatively easily extended to the similar problems of power-frequency overvoltages or to include more complex cases of multiconductor lines.

In the case of a nonuniform transmission line, at the point of discontinuity we have reflection and refraction of traveling waves and in such cases the problem becomes much more complex but can be solved by the method developed here. Also, in the case of lightning, the created overvoltages have a whole spectrum of very high frequencies. However, by using Fourier transform, an overvoltage wave of any shape can be decomposed in superposition of an infinite number of sinusoids, which constitute its spectrum [1]. Thus for an analysis of this type of overvoltage, the transient voltage of each of these frequencies can be separately considered. To do this, for each of these frequencies the relevant parameter of the considered line should be separately determined; this is also a complex problem, but can be solved by developing a corresponding computer program on the basis of the method presented here.

The developed analytical procedure successively follows, one by one, each phase of the actual physical process preceding the line steady state, by applying classical, well-known mathematics. As such, it is by itself very convenient for understanding and analyzing transient processes and resonant phenomena on power lines. By using relatively simple mathematical operations it is possible to determine transient and steady-state overvoltages and overcurrents, which, as is well known, can provoke different and frequently undesirable effects such as insulation damage, ground faults, malfunctioning of line protection devices, etc.

2.2 LONG GROUNDING CONDUCTORS

2.2.1 Introduction

Long grounding conductors are very important constitutive elements of the complex grounding systems of HV substations. As a rule their primary function is quite different, but in many cases their contribution to solving the grounding problem of these substations can be of prime importance. Therefore the contemporary conception of solving grounding problems in HV installations is based on the utilization of grounding effects of long grounding conductors: ground wires of overhead lines and metal sheaths of cable lines. Since they are constructed for the needs of their basic function, they exist independently of our need to solve grounding problems in HV substations. Thus, their grounding effects can be considered as certain grants that only should be taken as they actually are. According to this our interest is to know the characteristics of long grounding conductors and to utilize their grounding effects as much as possible. It is therefore justified to consider the relevant characteristics of long grounding conductors separately.

According to their design characteristics, all long grounding conductors can be divided into two groups. Those that are in direct and continuous contact with the earth, uncovered metal sheaths of cable lines, different types of metal pipelines, railways, etc., and those that are only indirectly and only at certain places in contact with the earth, ground wires of overhead lines and covered metal sheaths of cable lines in MV networks.

2.2.2 Uncovered Metal Sheath of Cable Lines

The uncovered metal sheath of cable lines as well as any of the long underground metal installations with direct and continuous contact with the earth, under the assumption that the surrounding soil is homogeneous, can be considered as one electrical line with specific parameters that are uniformly distributed along its length. By the discretization of these parameters on sections of the adopted length of 1 m, the metal sheath of a cable line of finite length can be presented by a corresponding lumped parameter ladder circuit, as shown in Fig 2.5.

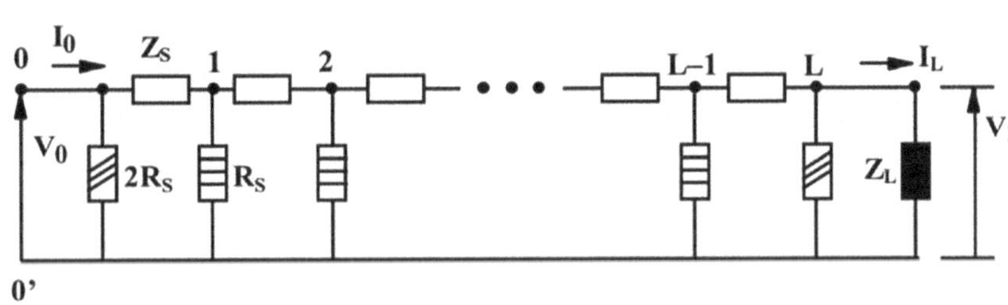

Figure 2.5 Equivalent circuit of cable line metal sheath [16].

The notation used has the following meaning:
- V_0 and I_0—potential and current at the beginning of the cable line sheath,
- V_L and I_L—potential and current at the end of the cable line sheath,
- R_S—grounding resistance of the metal sheath on the length of 1 m,
- Z_S—self-impedance of the metal sheath on the cable section of 1 m, and
- Z_L—grounding impedance at the end of the cable line.

As can be seen the lumped parameter ladder circuit of a line with distributed parameters represents a chain of identical π-quadruples, or involves the resistances $2R_S$ at the line ends. In a general case, i.e., for arbitrary values of Z_S, R_S, Z_L, and L, the potentials and currents at the ends of the circuit shown in Fig. 2.5 are, according to Eq. (2.45), related to:

$$\begin{aligned} &1.\ V_0 = \frac{k^L + k^{-L}}{2} V_L + \frac{(k^L - k^{-L}) Z_C}{2} I_L \\ &2.\ I_0 = \frac{k^L - k^{-L}}{2Z_C} V_L + \frac{k^L + k^{-L}}{2} I_L, \end{aligned} \tag{2.50}$$

where L is the relative cable line length expressed in relation to the length of 1 m and Z_C is the input (grounding) impedance of metal sheath of an infinitely long cable line.

The parameter k represents, according to the previous consideration, the "current distribution factor" in nodes 1, 2, 3, ..., assuming that the cable line is infinite ($L \to \infty$). This parameter is, according to Eq. (2.6), determined by:

$$k = 1 + \frac{Z_C}{R_S}, \tag{2.51}$$

where:

$$Z_C \approx Z_\infty = \frac{Z_S}{2} + \sqrt{R_S Z_S + \frac{Z_S^2}{4}}. \tag{2.52}$$

Since the adopted section of discretization is long only 1 m ($R_S >> |Z_S|$) instead of Eq. (2.52), for the determination of the grounding impedance of the cable line of infinite length, Z_C, we can use the following approximation:

$$Z_C \approx \sqrt{RZ_S}. \tag{2.53}$$

On the basis of this approximation, instead of Z_C we will now use $Z_{c\infty}$ as a symbol representing input (grounding) impedance of an infinite cable line.

Under the assumption that the cable metal sheath is an equipotential electrode over its entire length, the grounding resistance of the metal sheath on the cable line section of 1 m, according to Ref. [18], is:

$$R_S = \frac{\rho}{\pi m} \ln \frac{L}{\sqrt{h d_{sh}}}, \ \Omega, \tag{2.54}$$

where ρ is the equivalent soil resistivity along the cable line (Ωm), L is the cable line length, h is the depth at which the cable line is laid (m), d_{sh} is the outer diameter of the cable sheath (m), and m is the length of 1 m.

Since cable lines are used almost exclusively in urban areas the value of ρ can be only estimated on the basis of the main geological characteristics of the soil in which the cable is laid. So far, investigations of soil effects on the grounding impedance of long grounding conductors showed that the adoption of equivalent soil resistivity higher (up to 20%) than the one approximately estimated ensured results on the safe side (see Section 2.3).

Impedance Z_S is the self-impedance of the cable sheath on the length of 1 m, or $Z_S = Z'_{sh} \times 1$ m. This impedance is, according to Ref. [18], determined by:

$$Z'_{sh} = R'_{sh} + \omega \frac{\mu_0}{8} + j\omega \frac{\mu_0}{2\pi} \left(\frac{\mu_r}{4} + \ln \frac{\delta}{r_{sh}} \right), \ \Omega/\text{km}, \tag{2.55}$$

where R'_{sh} is the cable sheath resistance (Ω/km), r_{sh} is the medium radius of the cable sheath (m), ω is the angular frequency, $\omega = 2\pi f$, μ_r is the relative permeability of the metal sheath, μ_0 is the magnetic permeability of a vacuum, $4\pi \cdot 10^{-7}$ Vs/Am, and δ is the equivalent earth penetration depth (m).

The equivalent earth penetration depth is determined by:

$$\delta = 658 \sqrt{\frac{\rho}{f}}, \ \text{m}, \tag{2.56}$$

where ρ is the equivalent soil resistivity along and around the cable line (Ωm) and f is the power frequency, 50 or 60 Hz.

Here it should be mentioned that the analytical expressions (2.55) and (2.56) are based on Carson's theory of the ground fault current return path through earth [19–21].

2.2.3 Overhead Line Ground Wire(s)

When we consider the long grounding conductor formed by the transmission line ground wire and its connections with tower footing electrodes, the equivalent circuit is somewhat different. The electrical circuit formed through the ground wire(s) of the transmission line or one of its sections

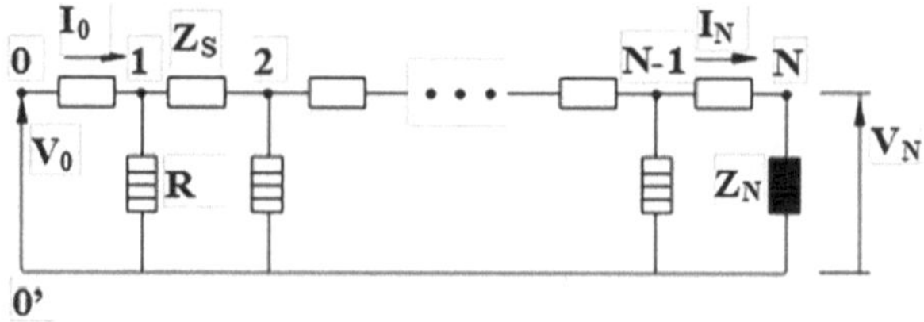

Figure 2.6 Equivalent circuit of an overhead line ground wire [15].

during ground fault with certain idealizations and approximations is presented in Fig. 2.6.

The notation used has the following meaning

- V_0 and I_0—potential and current at the beginning of the ground wire,
- V_N and I_N—potential and current at the end of the ground wire,
- R—average tower footing resistance,
- Z_S—self-impedance of the ground wire on one span of average length,
- Z_N—grounding impedance at the end of the line, and
- N—total number of spans.

As can be seen, the circuit consists of a finite number N of self-impedances of ground wire per one span, Z_S, and $(N-1)$ of resistances representing the average tower's footing resistance, R. This circuit is formed by introducing the following idealization of the actual physical model: the lengths of all the line spans are mutually equal and the tower's footing resistances are mutually equal. In this way, only those approximations we are forced to introduce are used since we cannot, because of the uncertainty of the basic data about soil resistivity, know at the design stage the exact values of each tower footing resistance. Also, tower footing resistances in practical conditions are strongly dependent on seasonal weather changes and cannot be treated as constant parameters. Thus for the purposes of calculation we are forced to adopt an approximately estimated value of this parameter.

In general, i.e., for any values of all presented parameters Z_S, R, N, and Z_N, the potentials and currents at the ends of the circuit are, according to Eq. (2.39), connected by the following equations:

$$
\begin{aligned}
&1.\ V_0 = \frac{k^N + k^{-N+1}}{1+k} V_N + \frac{(k^N - k^{-N})Z_\infty}{1+k} I_N, \\
&2.\ I_0 = \frac{k^N - k^{-N+2}}{(1+k)Z_\infty} V_N + \frac{k^N + k^{-N+1}}{1+k} I_N.
\end{aligned}
\tag{2.57}
$$

The current distribution parameter k is defined by:

$$k = 1 + \frac{Z_\infty}{R}, \tag{2.58}$$

where Z_∞ represents the grounding/input impedance of the ground wire(s) under the assumption that the line is infinitely long or that the number of spans is infinite ($N \to \infty$). This impedance is determined by:

$$Z_\infty = \frac{Z_S}{2} + \sqrt{RZ_S + \frac{Z_S^2}{4}}, \ \Omega. \tag{2.59}$$

The impedance Z_S is determined by:

$$Z_S = L_S Z'_{gw}, \ \Omega, \tag{2.60}$$

where L_S is the average length of the span (km) and Z'_{gw} is the self-impedance of the ground wire (Ω/km).

The self-impedance of the ground wire is, according to Ref. [22], determined by:

$$Z'_{gw} = R'_{gw} + \omega\frac{\mu_0}{8} + j\omega\frac{\mu_0}{2\pi}\left(\frac{\mu_r}{4} + \ln\frac{\delta}{r_{gw}}\right), \ \Omega/\text{km}, \tag{2.61}$$

where R'_{gw} is the ground wire(s) resistance (Ω/km), r_{gw} is the radius of the ground wire (m), and μ_r is the relative permeability of earth wire. The meaning of ω, μ_0, and δ is the same as in expression (2.55).

Eq. (2.57) can be used to describe voltages and currents of the ground wire(s) of the entire line or of any of its sections.

2.2.4 Simplified Equivalent Circuits

The possibility of simplifying indispensable calculations is seen in the reduction of the number of elements of the equivalent circuits presented in Figs. 2.5 and 2.6. Thus by using the system of Eq. (2.50), one can determine the analytical expressions for the elements of the one equivalent π-cell shown in Fig. 2.7, substituting the whole circuit in Fig. 2.5. Also by using

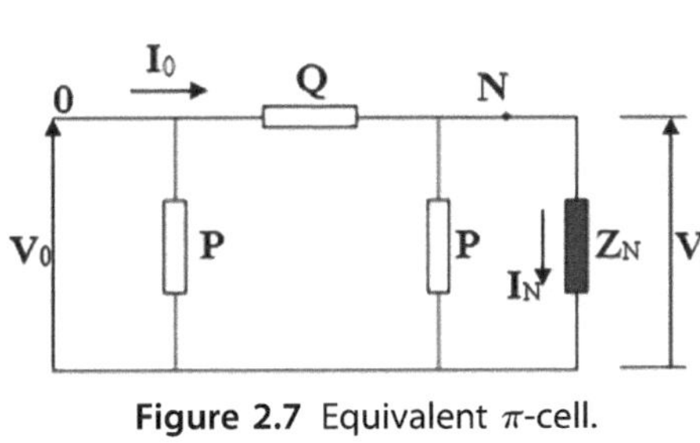

Figure 2.7 Equivalent π-cell.

the system of Eq. (2.57), one can determine the analytical expressions for the elements of the one equivalent π-cell, Fig. 2.7, substituting the whole circuit presented in Fig. 2.6.

According to Kirchhoff's rules, the voltages and currents at the ends of the circuit shown in Fig. 2.7 are connected with the following system of equations:

$$\begin{aligned} &1.\ V_0 = \left(1 + \frac{Q}{P}\right) V_N + Q I_N \\ &2.\ I_0 = \left(\frac{Q+P}{P^2}\right) V_N + \left(1 + \frac{Q}{P}\right) I_N \end{aligned} \tag{2.62}$$

To replace the equivalent circuit from Fig. 2.5 with the circuit shown in Fig. 2.7, the voltages and currents at the ends of the circuit in Fig. 2.7 must be equal to the corresponding voltages and currents of the circuit in Fig. 2.5. This means that the coefficients of the system of Eq. (2.62) must be equal to the corresponding coefficients of Eq. (2.57). By making these coefficients equal we obtain a system of equations whose solution gives the following analytical expressions:

$$Q_{sh} = \frac{k^L - k^{-L}}{2} Z_{c\infty}, \tag{2.63}$$

$$P_{sh} = \frac{k^L + 1}{k^L - 1} Z_{c\infty}. \tag{2.64}$$

An analogous procedure gives the analytical expressions determining the equivalent impedances for the simple circuit (Fig. 2.7) representing transmission line ground wire(s) as a grounding conductor:

$$Q_{gw} = \frac{k^N - k^{-N}}{1 + k} Z_\infty, \tag{2.65}$$

$$P_{gw} = \frac{k^N - k^{-N}}{k^N + k^{-N+1} - k - 1} Z_\infty. \tag{2.66}$$

Also for solving certain practical problems instead of impedances Q_{sh} and P_{sh} it is more useful to know the input/grounding and transfer/axial impedance of the lumped parameter ladder circuit presented in Fig. 2.5. On the basis of Eq. (2.50) these impedances are determined by:

$$Z_{gsh} = \frac{U_0}{I_0} = \frac{(k^L + k^{-L}) Z_L Z_{c\infty} + (k^L - k^{-L}) Z_{c\infty}^2}{(k^L - k^{-L}) Z_L + (k^L + k^{-L}) Z_{c\infty}}, \tag{2.67}$$

$$Z_{0L} = \frac{U_0}{I_L} = \frac{(k^L + k^{-L}) Z_L + (k^L - k^{-L}) Z_{c\infty}}{2}. \tag{2.68}$$

On the basis of Eq. (2.68), any lumped parameter ladder circuit shown in Fig. 2.5 can be, from the standpoint of U_0, U_L, and I_L, substituted by the very simple equivalent circuit presented in Fig. 2.3. This certainly significantly reduces the calculations necessary for the determination of U_L and I_L that in solving practical problems usually appear as unknown quantities.

Analogously to this, on the basis of Eq. (2.61), the input and transfer impedances of the circuit representing ground wire(s) are:

$$Z_{ggw} = \frac{U_0}{I_0} = \frac{(k^N + k^{-N+1})Z_N Z_\infty + (k^N - k^{-N})Z_\infty^2}{(k^N - k^{-N+2})Z_N + (k^N + k^{-N+1})Z_\infty}, \quad (2.69)$$

$$Z_{0N} = \frac{U_0}{I_N} = \frac{(k^N + k^{-N+1})Z_N + (k^N - k^{-N})Z_\infty}{1 + k}. \quad (2.70)$$

On the basis of Eq. (2.74), any lumped parameter ladder circuit in Fig. 2.6 can be, from the standpoint of U_0, U_N, and I_N, substituted by the very simple one presented in Fig. 2.3. This certainly reduces the calculations necessary for the determination of U_N and I_N.

The presented analytical expressions can be used for a number of practical problems such as determination: the distribution of the ground fault current for faults anywhere along the cable and overhead lines, as well as the potentials that can appear at a certain place along these lines.

2.2.5 Active Length of a Ground Wire as Grounding Conductor

In determining grounding effects of transmission line ground wire(s) the necessary calculation procedure can be reduced if we know the active ground wire length or the length on which a certain transmission line ground wire performs its grounding function. In this case, instead of the relatively complicated analytical expression (2.69) for determining this grounding impedance the simpler expression (2.59) can be used. However, how can we know whether the certain transmission line is infinitely long from the standpoint of grounding effects?

First, we shall assume that we have a transmission line and that its length is infinite, i.e., that its end is somewhere in infinity. Then, we shall imagine that this line is divided into two sections: one from the beginning to an arbitrary point N along the line and the other from point N to infinity. Under such assumptions the circuit formed by the ground wire of this line can be represented by the simple equivalent circuit shown in Fig. 2.8.

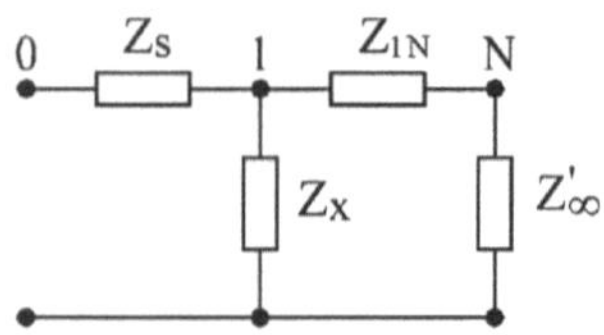

Figure 2.8 Simple equivalent circuit of an infinite line ground wire [15].

The used notation has the following meaning:

- Z_X—grounding impedance of the ground wire seen from the first to the Nth tower,
- Z_{1N}—transfer impedance of the ground wire from the first tower to the Nth tower,
- Z'_∞—grounding impedance of the ground wire(s) seen from any tower (Nth) toward the line end, and
- N—number of spans of the first line section.

As it has been assumed that the line is infinitely long, the grounding impedance of the ground wire seen from any tower along the line toward its end has the same value. According to the circuit in Fig. 2.6 and expression (2.52) this impedance is:

$$Z'_\infty = Z_\infty - Z_S = \sqrt{RZ_S + \frac{Z_S^2}{4}} - \frac{Z_S}{2}. \tag{2.71}$$

Then, on the basis of Eq. (2.71) and the circuit in Fig. 2.8, we can write the following relation:

$$Z'_\infty = \frac{Z_X\left(Z_{1N} + Z'_\infty\right)}{Z_X + Z_{1N} + Z'_\infty}, \tag{2.72}$$

or:

$$Z_X = \left(1 + \frac{Z'_\infty}{Z_{1N}}\right) Z'_\infty. \tag{2.73}$$

According to Eq. (2.12), transfer impedance Z_{1N} is determined by:

$$Z_{1N} = \frac{k^{N-1} - 1}{k - 1} Z_S, \tag{2.74}$$

or, on the basis of Eq. (2.58):

$$Z_{1N} = \frac{(k^{N-1} - 1)RZ_S}{Z_\infty}. \tag{2.75}$$

Finally, as Eqs. (2.59) and (2.71) give $RZ_S = Z_\infty Z'_\infty$, we have:

$$Z_X = (k^{N-1} - 1)Z'_\infty. \tag{2.76}$$

By using Eq. (2.76), expression (2.73) becomes:

$$Z_X = \left(1 + \frac{1}{k^{N-1} - 1}\right)Z'_\infty. \tag{2.77}$$

When the length of the first section is approximately equal to the active line length, N_a, then the grounding impedance Z_X must be approximately equal to the impedance Z'_∞. According to expression (2.77), this condition can be expressed by the following inequality:

$$\frac{1}{k^{N-1} - 1} \leq \frac{1}{k^{N_a-1} - 1} = \varepsilon, \tag{2.78}$$

where ε is the relative error in the value of impedance Z_X: $|Z_\infty - Z_X|/|Z_\infty|$, an arbitrarily small, a priori adopted number, and N_a is the active line length, from the standpoint of grounding effects, expressed in number of spans.

When we add the first line span to the number of spans $(N - 1)$ the explicit form of Eq. (2.78) becomes:

$$N \geq N_a = \frac{\ln\left(1 + \frac{1}{\varepsilon}\right)}{\ln\left(1 + \sqrt{\frac{|Z_S|}{R}}\right)}. \tag{2.79}$$

Since for practical purposes a sufficiently small ε can be considered, 0.05, instead of Eq. (2.79) we can use:

$$N \geq N_a(\varepsilon = 0.05) \approx \frac{3}{\ln\left(1 + \sqrt{\frac{|Z_S|}{R}}\right)}. \tag{2.80}$$

The ground wires of the transmission lines are the most numerous elements of grounding systems belonging to HV transmission and power stations. The actual length of the transmission lines outgoing from these substations is such that almost without exception the criterion defined by Eq. (2.80) is satisfied. Thus the grounding contribution of these grounding system elements can be determined by using expression (2.59). In the relatively rare cases when supply and supplied substation are located at mutually small distances, i.e., less than the active ground wire length, expression (2.69) is more accurate.

2.2.6 Active Length of the Cable Line as Grounding Conductor

It is well known that MV paper-insulated cable lines have uncoated metal sheath and therefore act as long grounding electrodes or dissipate fault current directly into the surrounding soil. As a consequence the MV lines for this type of cable spontaneously form a very large and complex grounding system around the HV/MV distribution substations. The spontaneously formed grounding systems manifest excellent characteristics, including a very low value of impedance to ground. However, the problem arises when this favorable fact is confirmed by calculations. Thus the problem with this type of grounding systems appears at the design stage of an HV/MV substation, when safety conditions inside and in the vicinity of this substation should be estimated and conformed. Therefore it is necessary to have at one's disposal an appropriate and sufficiently accurate calculation method.

The grounding system of the HV/MV distribution substations is not strictly limited to an a priori defined area. They are formed freely and spontaneously so that they encompass various underground metal installations and structures, including those that only trivially, because of the great distance from the HV/MV substation, perform a grounding function, i.e., dissipate ground fault current into the earth. Thus it can be said that the space limits of the grounding system formed in this way are practically identical to the space limits of the potential field produced by the grounding system itself. Certainly, the area encompassed in this way is the largest possible and is determined by the active length of the outgoing MV cable lines seen from the considered HV/MV substation. As a consequence the grounding systems spontaneously formed around the HV/MV substations have characteristics that are extremely favorable for the certain design characteristics of the implemented MV cables and for the certain geological characteristics of the encompassed urban area [23,24]. Thus the first question that should be asked is: how can we estimate the size of the area covered by such a grounding system?

In the most number of cases two to three tens of cable lines leave an HV/MV distribution substation and cover the whole area around this substation in all available directions. Thus the realistic physical model (electrical circuit) of spontaneously formed grounding systems has a very complex structure and configuration, which differ from case to case depending on the local urban surroundings in each concrete case. As a result, certain idealizations and approximations of the actual physical model should be treated as unavoidable. On the basis of such an approach the area covered by a spontaneously formed grounding system can be defined as an approximately circular surface with the center in the HV/MV substation itself and with the radius approximately equal to the active length of the outgoing MV cable lines. This means that when the active length of the outgoing cable lines for an HV/MV substation is determined, the urban area covered by its grounding system is also approximately determined.

In contemporary power distribution practice the cable lines with uncoated metal sheaths are usually realized through an armored lead-sheathed three-core MV cable. The metal sheath of such an MV cable line represents an electrical line with specific parameters that are, if we assume homogeneous surrounding soil, uniformly distributed along its length. By the discretization of these parameters on sections of length of 1 m, the metal sheath of the cable line of finite length can be represented by the equivalent circuit shown in Fig. 2.5. Then, the voltages and currents of the metal sheath at the cable line ends are, in general, mutually connected with the system of Eq. (2.50).

In determining the active cable line length, we shall start, as in the case of overhead line ground wire, from an imagined case distant from practice, but very convenient for analysis. Namely, we shall assume that the cable line is infinitely long or that its end is somewhere in infinity. Then, we shall assume that this cable line is fictitiously divided into two sections: one from the beginning to an arbitrary point *X* along the line and the other from this point to infinity. On the basis of previous assumptions, a cable line as a grounding conductor can be represented by the simple equivalent circuit shown in Fig. 2.9.

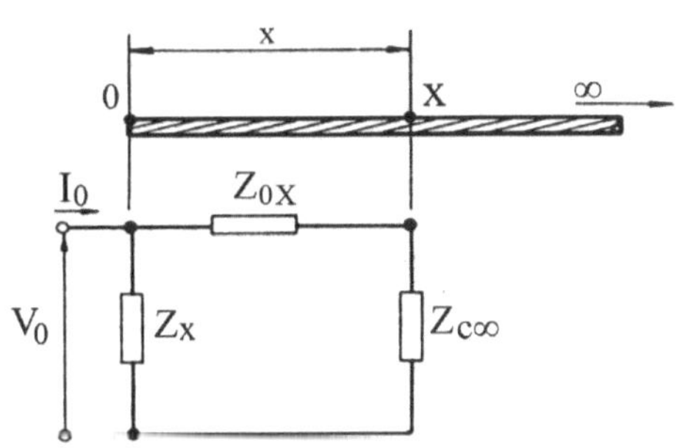

Figure 2.9 Equivalent circuit of an infinite cable line metal sheath [23].

On the basis of the given equivalent circuit, the grounding impedance of the cable line metal sheath can be presented as:

$$Z_{c\infty} = \frac{(Z_{0X} + Z_{c\infty})Z_x}{Z_x + Z_{0X} + Z_{c\infty}}, \tag{2.81}$$

while the grounding impedance of the first cable line section is:

$$Z_x = \left(1 + \frac{Z_{c\infty}}{Z_X}\right) Z_{c\infty}. \tag{2.82}$$

By applying Eq. (2.50) on the first section of the assumed line, it can be shown that transfer/axial impedance is given by:

$$Z_{0X} = \frac{V_0}{I_X} - Z_{c\infty} = (k^x - 1)Z_{c\infty}, \tag{2.83}$$

where x is the relative length of the first section expressed in relation to the length of 1 m.

Where the length of the first section is so large that it can be considered as infinitely long, then grounding impedance Z_X approaches near to impedance $Z_{c\infty}$. This criterion is fulfilled if, according to Eqs. (2.54), (2.55), and (2.82), the following inequality is satisfied:

$$x \geq L'_a = \frac{\ln(1 + 1/\varepsilon)}{\ln\left(1 + \sqrt{\dfrac{|Z_S|}{R(L_a)}}\right)}, \tag{2.84}$$

where L'_a is the relative value of the active cable line length expressed in relation to the length of 1 m, ε is the relative error in the value of impedance Z_X: $|Z_{c\infty} - Z_X|/|Z_{c\infty}|$, a prior adopted arbitrarily small number, usually 0.05.

Then, by introducing the expression for $R(L_a)$ in Eq. (2.84) we obtain:

$$x \geq L'_a = \frac{\ln(1 + 1/\varepsilon)}{\ln\left(1 + \sqrt{\dfrac{\pi|Z_S|m}{\rho \ln \frac{L_a}{\sqrt{hd}}}}\right)}. \tag{2.85}$$

Eq. (2.85) represents a transcendental equation in which the only unknown quantity is the active cable line length. Such an equation may be solved graphically by representing the left and right sides of this equation as two separate functions of L_a or by several iterative calculations.

On the basis of the presented derivation procedure, it is clear that the cable sheath section of any length at a distance greater than L_a, as well as the associated terminal grounding electrode(s) of any magnitude, practically does not perform a ground function or does not dissipate ground fault current into the earth. In other words, any increase in the length of a cable line over its active length has no noteworthy influence on the value of cable line grounding impedance, viewed from its beginning. Also this cable line length can be defined as a length for which transfer impedance of the metal sheath (Eq. 2.68) becomes practically infinitely large, or current, I_L, practically equal to zero (Fig. 2.5). However, expression (2.58) is determined under the assumption that the potential appearing on a long grounding conductor is approximately equal along its entire length. Since this is obviously not the case when a cable line can be treated as infinitely long, it is necessary to consider the influence of this approximation on the accuracy of the derived Eq. (2.85).

2.2.7 Influence of Nonequipotentiality of the Metal Sheath

MV cable lines, outgoing from a distribution HV/MV substation, usually supply in series several, mostly 10 MV/LV, substations, and, in typical urban areas, have a length that is most frequently in the range of 3−5 km. On the basis of this fact and expression (2.84), it is not difficult to show that all of them are significantly longer than their active length. Thus all MV cable lines outgoing from a distribution HV/MV substation can be treated, from the standpoint of grounding effects, as infinitely long. This means that the potentials appearing during a ground fault on the metal sheath of the cable lines and on the substation grounding electrode are mutually equal only in the HV/MV substation itself. Outside the HV substation, the metal sheath potential decreases with an increase in the distance from this HV/MV substation. At distances that are equal to or larger than the active cable line length, the potential of the metal sheath becomes practically equal to the potential of the surrounding earth or to zero potential. Thus cable lines of greater distance cannot be more active in ground fault current dissipation into the surrounding earth. Since expression (2.54) has been derived assuming equipotentiality over the entire cable line length, its application in considered cases is not quite correct or introduces a certain inaccuracy. It is therefore necessary to see the influence of this approximation on the accuracy of the formerly derived expression (2.85).

Distribution of the sheath potential along the cable line that can be treated as infinitely long ($L \to \infty$) is, according to Eq. (2.50), given by:

$$V_X = V_0 k^{-x}, \tag{2.86}$$

where x is the distance to the arbitrary point along the cable line, expressed in relation to the length of 1 m.

Here it should be mentioned that during a ground fault the substation grounding electrode forms a very large potential field around the area covered by the HV/MV substation. However, this fact has no practical influence on the potential distribution along the cable line that is infinitely long [23,24].

Since the actual cable line lengths are longer than even two active cable line lengths, each actual MV distribution cable line, viewed from any point along the section on which the cable line performs its grounding function, can be treated as infinitely long. On the basis of the foregoing, the current that dissipated into the earth along such cable line is, according to Eq. (2.50), distributed in the following manner:

$$I_X = \frac{V_0}{R(L_a)} k^{-x} = I_1 k^{-x}, \tag{2.87}$$

where I_X is the current dissipated into the earth from the arbitrary cable section that is on the potential denoted as V_X, $R(L_a)$ is the grounding resistance of the cable section of 1 m when the cable line length is equal to the active cable line length, and I_1 is the current dissipated into the earth from the first (nearest) cable segment of 1 m length. A graphical presentation of the intensity of the current dissipated into the earth along the cable line of infinite length is presented in Fig. 2.10.

The total current dissipated into the earth is, according to the equivalent circuit in Fig. 2.5, equal to I_0, so that one can write the following relation:

$$I_0 = I_1 \int_0^{\infty} k^{-x} dx = I_1 \sqrt{\frac{R(L_a)}{Z_S}} \tag{2.88}$$

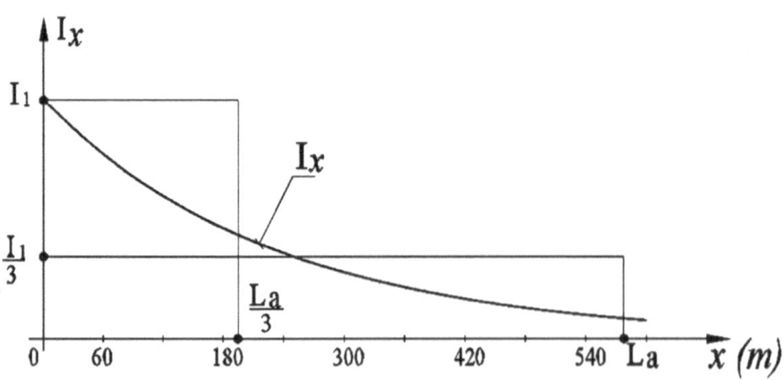

Figure 2.10 Variation of current I_X along the cable line of infinite length [18].

We will now make an unrealistic assumption that current I_0 is dissipated into the earth without any attenuation over the entire cable line length. The length of this fictitious cable line necessary for dissipation of the current I_0 is, according to Eq. (2.88), determined by:

$$L_e' = \frac{I_0}{I_1} = \sqrt{\frac{R(L_a)}{Z_S}}, \tag{2.89}$$

where L_e' is the relative length of the fictitious cable line with the constant sheath potential over entire its length, expressed in relation to the length of 1 m. By using expressions (2.84) and (2.88) it is possible to establish the following relation:

$$L_a' = \frac{\ln(1 + 1/\varepsilon)}{\ln\left(1 + 1/L_e'\right)}, \tag{2.90}$$

or, in explicit form:

$$L_e' = \left(e^{\frac{\ln(1+1/\varepsilon)}{L_a'}} - 1\right)^{-1}. \tag{2.91}$$

On the basis of the quantitative analysis performed by using Eq. (2.90), it is not difficult to show that L_e and L_a are mutually connected with the following simple relation:

$$L_e = \frac{L_a}{3}. \tag{2.92}$$

Finally, the grounding resistance of 1 m of this fictitious cable line is, according to Eqs. (2.54) and (2.92), given by:

$$R(L_e) = \frac{\rho}{\pi m} \ln \frac{L_a}{3\sqrt{hd}}. \tag{2.93}$$

Obviously, two different values for grounding resistance of the metal sheath of 1 m, $R(L_a)$ and $R(L_a/3)$, correspond to two theoretically extreme cases from the standpoint of current dissipation into the surrounding earth, shown in Fig. 2.10 by straight lines. According to one of them, the current that dissipated into the earth has a constant value over the entire active cable line length, while according to the other its value is constant over the first third of the active cable line length, and it is equal to zero on the rest of the cable line. Thus it is clear that an actual value of the mentioned resistance should be somewhere between $R(L_a)$ and $R(L_a/3)$. However, a favorable circumstance is that the active cable line length is only slightly dependent upon the value of this resistance, which can be seen on the basis of expression (2.84). By using this expression, it is not difficult to show that the value of the active cable line length obtained by using $R(L_a/3)$ is, for a wide range of possible soil resistivities, only from 5% to 8% less than the one obtained by using $R(L_a)$.

On the basis of the former consideration it is clear that the influence of metal sheath nonequipotentiality along an infinite cable line has only a slight influence on the active cable line length. Bearing this in mind, for the active cable line length the following approximation can be used:

$$L_a(\varepsilon = 0.05) \approx \frac{3m}{\ln\left(1 + \sqrt{\dfrac{\pi |Z_S| m}{\rho \ln \frac{L_a}{\sqrt{hd}}}}\right)}, \tag{2.94}$$

where L_a is the active cable line length in meters (m) and m is the length of 1 m.

The given expression is derived under the assumption that the considered line is infinitely long, even looking at it from the end of the section that is equal to its active length. This means that expression (2.94) is mostly accurate in cases of cable lines when their physical length is 2 times greater than their active length. Since cable lines in the MV distribution network are usually several kilometers long, Eq. (2.94) can be used in almost all practical situations. The given expression can also be used for other types of long grounding conductors such as different metal pipelines that are in direct and continuous contact with the earth.

As has been already mentioned the urban area covered by the grounding system of an HV substation can be described as an approximately circular surface around this substation with a radius equal to the active length of outgoing MV cable lines. As a result, as will be seen later in Chapter 6, this length has a very important role in the procedure of developing the analytical method for determining the main characteristics of this type of grounding system.

2.3 CHARACTERISTICS OF LONG GROUNDING CONDUCTORS

2.3.1 The Influence of Soil Resistivity

The contemporary conception of solving grounding problems in HV substations is based on all available recourses. First of all the grounding effects of long grounding conductors: grounding wire(s) of overhead lines, metal sheath(s) of cable lines, and different types of metal pipelines. Since these conductors exist independently of our need to solve the mentioned problem this conception enables the most economical solution in each concrete case. As a result, the grounding effects of this type of grounding conductor deserve to be separately considered and completely clarified. One more reason for this is also very important and emanates from the fact that these conductors are in the greatest part situated outside of HV installation fences and in some public places may be a potential source of danger to humans and other creatures.

With the aim of obtaining a better insight into the specific particularities of this type of grounding conductor, instead of the usual equivalent circuit of a transmission line ground wire presented in Fig. 2.6 we will consider the especially adjusted equivalent circuit presented in Fig. 2.11.

The used notation has the following meaning:

- Z_1, Z_2, …, Z_n, …, Z_N—equivalent impedances between the line beginning, 0, and each of the tower footing electrodes seen separately,
- Z_g—grounding impedance of the substation at the line end, and
- R—average tower footing resistance.

On the basis of the previously derived analytical expressions for input and transfer impedances of the lumped parameter ladder circuits, the circuit presented in Fig. 2.11 can be treated as equivalent to the circuit in Fig. 2.6, from the standpoint of the current injected into the surrounding earth at any (nth) of the points (towers), if impedance Z_n is determined by:

$$Z_n = \frac{R + Z_{gn}}{Z_{gn}} Z_{0n}, \tag{2.95}$$

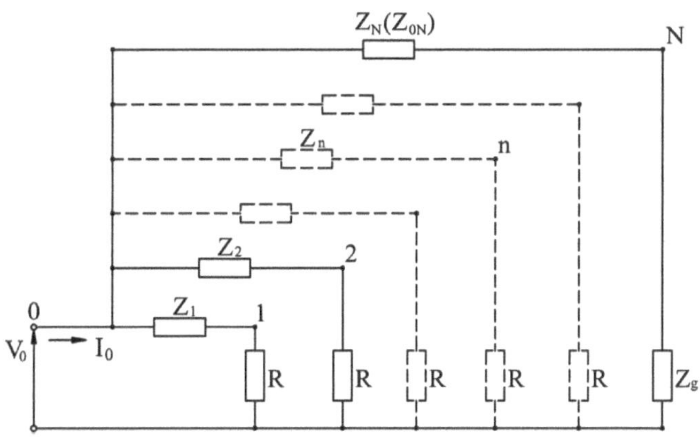

Figure 2.11 Equivalent circuit with tower footing resistances individually connected [7].

where Z_n is the impedance between the beginning of the line and the nth tower considered separately, Z_{gn} is the grounding impedance of the ground wire seen from the nth tower toward the line end, and Z_{0n} is the transfer impedance between the beginning of the line and the nth tower.

Impedances Z_{gn} and Z_{0n} are determined by the previously derived Eqs. (2.69) and (2.70), respectively. Since impedance Z_{gn} seen from the line end is infinitely large ($Z_{gN} = \infty$), according to Eq. (2.95), impedance Z_N is equal to impedance Z_{0N}.

As can be seen, the equivalent circuit presented in Fig. 2.11 enables us to separately consider and see the participation of each of the tower footing electrodes when current I_0 is injected into the earth. On the basis of such insight, one can explain and understand the specific particularities of long grounding conductors.

As is well known, soil resistivity depends on the geological structure of the soil and depending on the type of soil can have any value in a very wide range from 10 to 10^4 Ωm [25]. As a result, only the boring test samples and other geological investigations provide necessary information on the presence of various layers and the nature of soil material at a certain site. However, for the purposes of the quantitative analysis that will be presented here it is sufficient to know that tower footing resistance increases linearly with the increase in soil resistivity.

2.3.2 Quantitative Analysis

The purpose of this analysis is to obtain the answer to the following question: how does soil resistivity influence the grounding impedance of a long grounding conductor? To obtain the answer to this question we will consider a practical example by adopting a transmission line ground wire with the following relevant data:

- Line ground wire: Al/steel 95/55 mm^2
- Total line length: 10 spans
- Average span length: 250 m

Impedance Z_g represents the grounding impedance of a nearby HV substation and its value in practical conditions is usually less than 0.1 Ω. Bearing in mind the actual values of all other impedances and resistances of the circuit in Fig. 2.6, it means that the influence of impedance Z_g on the grounding impedance of a transmission line ground wire can be disregarded, i.e., $Z_g \approx 0$. Then, according to the analytical expressions (2 64) and (2 65), we can use for Z_{gn} and Z_{0n} the following expressions:

$$Z_{gn} = \frac{k^{(N-n)} - k^{-(N-n)}}{k^{(N-n)} + k^{-(N-n)+1}} Z_\infty, \tag{2.96}$$

$$Z_{0n} = \frac{\left(k^n - k^{-n}\right) Z_\infty}{1 + k}. \tag{2.97}$$

For a better insight into the considered effects we will assume two identically constructed HV lines installed in areas with different soil resistivity. In one case the surrounding area has favorable soil resistivity of 20 Ωm, whereas the other area has a relatively high soil resistivity of 100 Ω/m. The corresponding values of the tower footing resistances, R, of these lines are equal: 3 and 15 Ω.

Also by using the given data and the presented analytical expressions, we can calculate the values of impedances Z_n ($n = 1, 2, 3, \ldots, 9$) and Z_N ($N = 10$) in Fig. 2.11, for both adopted lines. Then, we will determine these impedances under the assumption that the value of equivalent soil resistivity in both of the considered cases is doubled as a consequence of a long droughty period of weather. The results obtained for all impedances in all assumed cases are presented in Table 2.1.

As can be seen the values of these impedances increase when the values of soil resistivity become 2 times higher, but not two times, whereas the transfer impedance Z_N becomes even lower when soil resistivity becomes

Table 2.1 Effective value of impedance Z_n

	$\|Z_n\|$ (Ω)			
n	**20 Ωm**	**40 Ωm**	**100 Ωm**	**200 Ωm**
1	0.84	1.17	2.10	3.72
2	1.70	2.40	4.52	8.19
3	2.62	3.78	7.45	13.78
4	3.68	5.46	11.25	21.15
5	5.05	7.74	16.51	31.42
6	7.09	11.19	24.44	46.85
7	10.66	17.12	37.80	72.59
8	18.43	29.48	64.92	124.73
9	43.55	78.98	141.39	281.85
10	3.53	2.70	2.35	2.29

higher. Obviously, this is the reason why the grounding impedance of the entire transmission line ground wire, Z_{g0}, is not doubled, as is the case with tower footing resistances and all other grounding electrodes whose dimensions are such that potential appearing on them has approximately the same value at each point.

The relative changes in the value of impedances Z_n, ($n = 1, 2, 3, \ldots, 10$) when soil resistivity is doubled are shown in Fig. 2.12.

The decrease in the transfer impedance Z_{0N} (Z_N in Fig. 2.11) points to the fact that the influence of the ground wire, when connected with the grounding electrode at the line end, in cases of higher value of soil resistivity is intensified. Changes to this impedance for the different total line lengths and different soil resistivities are shown in Fig. 2.13.

As can be seen, for the same line length the transfer/axial impedance is lower when soil resistivity or tower footing resistance is higher. Obviously,

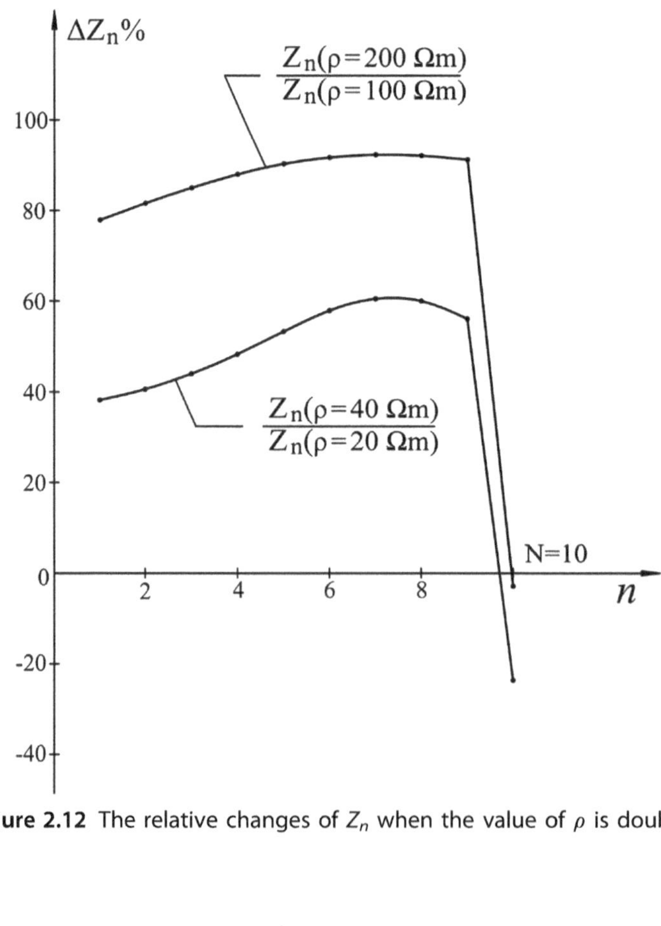

Figure 2.12 The relative changes of Z_n when the value of ρ is doubled.

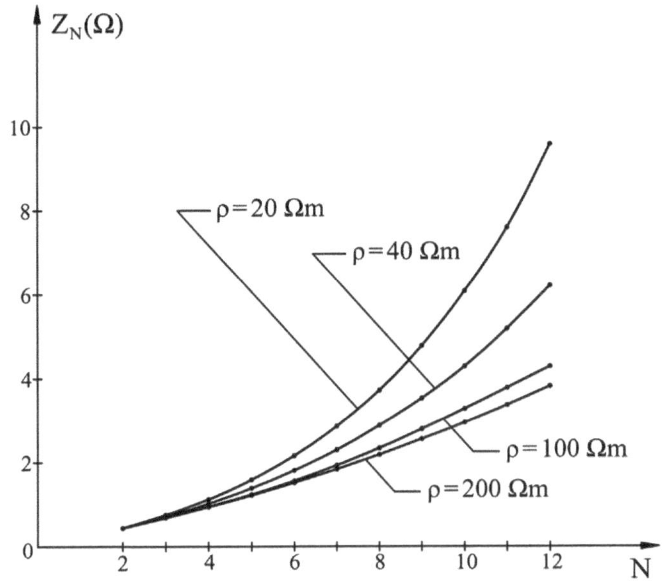

Figure 2.13 Impedance Z_N(i.e., Z_{0N}) as a function of line length and for different soil resistivities.

if we assume that the resistance of the tower footings tends to infinity this impedance obtains its minimal possible value equal to the self-impedance of ground wire of the whole line, $Z_{0N} = NZ_S$ or becomes a linear function of the line length. This obvious fact can be proven mathematically. In the case where the ground wire is closed with impedance $Z_g = Z_\infty - Z_S$, we can, according to Eq. (2.12), express impedance Z_{0N} immediately through impedance Z_S. Then, if we assume that the tower footing resistance, R, tends to infinity, according to Eq. (2.58), the value of parameter k tends to be equal to one ($k \rightarrow 1$) and we have:

$$\lim_{k\rightarrow 1} Z_{0N}(Z_g = Z_\infty - Z_S) = \frac{\frac{d(k^N - 1)}{dk}}{\frac{d(k-1)}{dk}} Z_S = NZ_S \tag{2.98}$$

Also, with the aim of better insight into the considered effect, the change of transfer/axial impedance to any tower along the considered lines, Z_{0n} ($n = 1, 2, 3, \ldots, 10$) is graphically presented in Fig. 2.14.

As can be seen, any increase in soil resistivity (or R) results in a certain decrease in the transfer/axial impedance of transmission line ground wire to

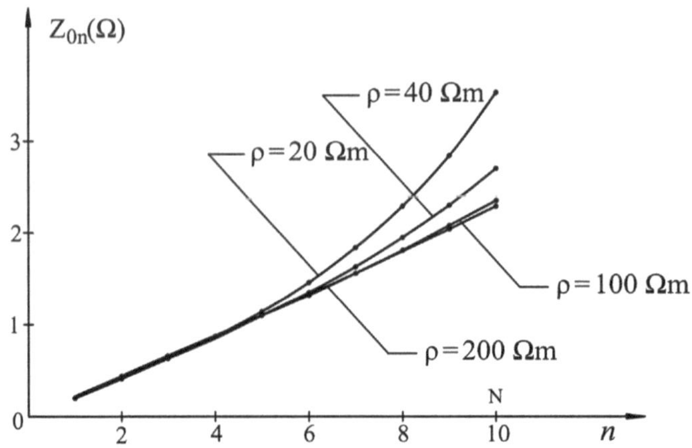

Figure 2.14 Impedance Z_{0n} as a function of line length and for different soil resistivities.

any point/tower along the line. This obviously intensifies grounding function of the more distant towers and compensates to a certain extent the unfavorable influence of high soil resistivity. This is certainly a common and very useful characteristic of all types of long grounding conductors.

Formerly performed analysis gives the answer to the question: why is the grounding impedance of the transmission line ground wire or in general of any long grounding conductor not strongly (linearly) dependent on soil resistivity? The variations of this impedance depending on line length expressed through the number of spans, n, and for extremely different values of the grounding impedance at the line end ($Z_g = 0$ and $Z_g = \infty$) in the case of the line with the tower footing resistance, $R = 6\ \Omega$, are given in Fig. 2.15.

On the basis of the given diagrams it is not difficult to see that for a certain line length, impedance Z_{g0} becomes constant and does not depend on a further increase in line length. At this length, which certainly represents the active ground wire length, impedance Z_{g0} becomes equal to impedance Z_∞, and its value for longer lines does not dependent on the value of the grounding impedance at the line end, Z_g. The way in which the active ground wire length changes with the increase in tower footing resistance can be considered by using expression (2.79). The results of calculations for the different types of ground wires and for the different desired degrees of accuracy, ε, are presented in Fig. 2.16.

The way in which the active cable line length and grounding impedance of infinitely long cable line, $Z_{c\infty}$, changes with the increase in soil

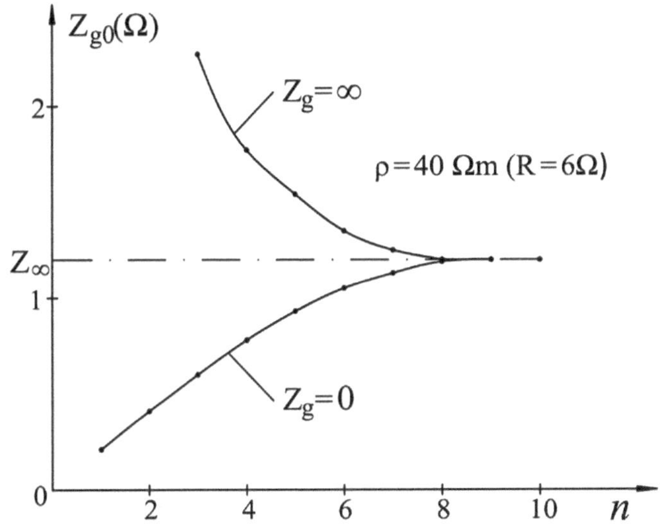

Figure 2.15 Variations of grounding impedance, Z_{g0} depending on line length.

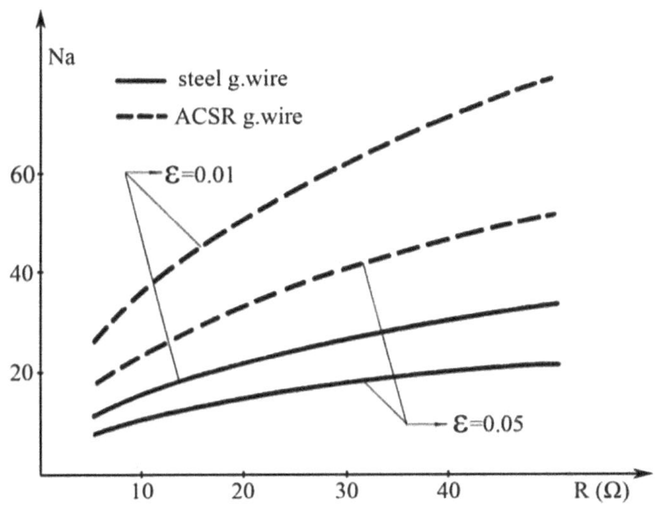

Figure 2.16 Active ground wire length depending on *R* [16]. *ACSR,* aluminum conductor steel reinforced.

resistivity can be seen in the example of an MV cable line performed by a cable of standard design characteristics and with an uncovered metal sheath. The results of calculations performed by using Eq. (2.94) are graphically presented in Fig. 2.17.

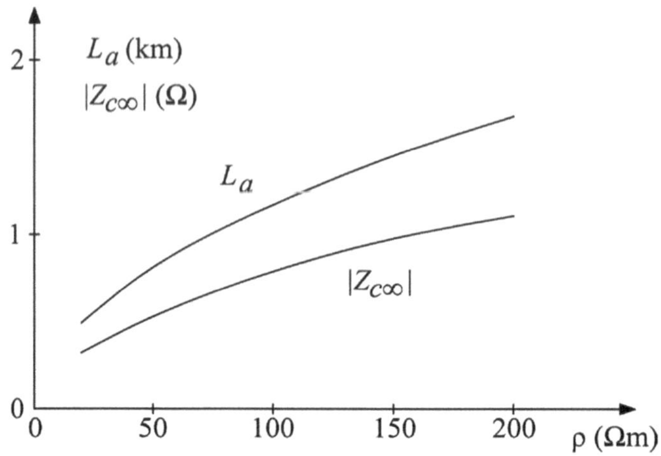

Figure 2.17 Variation of L_a and $|Z_{c\infty}|$ of a cable line depending on ρ [23].

The diagrams in Fig. 2.17 show the influence of the soil resistivity on the active cable line length and on the grounding impedance of the infinitely long cable line. On the basis of the diagram for the active cable line length, it is not difficult to see that the active cable line length depends to a great extent on the surrounding soil resistivity. The active cable line length spontaneously adapts its magnitude to the surrounding soil resistivity in such a manner that it is greater when the value of surrounding soil resistivity is greater. Also the grounding system formed from this type of MV cable lines around an urban HV/MV substation covers a larger area when natural conditions for solving grounding problems are less favorable, i.e., when the value of soil resistivity is higher. In such a way these grounding systems to some extent compensate for the unfavorable influence of natural factor, i.e., high soil resistivity [25].

On the basis of the presented diagram for the grounding impedance of the cable line of infinite length, it can be seen that this impedance is not strongly (linearly) dependent on soil resistivity, as in the case of the sufficiently small ground electrode that can be treated as equipotential on its entire surface. As a result, long grounding conductors of this type of MV cable line partially compensate for the unfavorable influence of high soil resistivity. The same is true in the case of increased soil resistivity, which can appear, as is well known, as a consequence of a long lasting period of dry weather.

In the case of a cable line that can be considered as infinitely long ($L \geq L_a$), beside smaller values of transferaxial impedances, Z_{0n} ($n = 1, 2, 3, \ldots N$), one more reason exists for the mentioned compensation. This is the fact that the active cable line length becomes somewhat greater, as can be seen from Fig. 2.17. With the aim of better insight into the considered effect, changes in transfer/axial impedance along a cable line for different values of soil resistivity are presented in Fig. 2.18.

This favorable particularity of long grounding conductors means at the same time that the relative participation of cable lines in dissipation of the fault current through the substation grounding system of an HV substation is greater, when the value of soil resistivity is higher, i.e., less favorable.

Also previous considerations implicitly show that the unavoidable uncertainty in the determination of equivalent soil resistivity in urban areas cannot have too large an influence on the value of the grounding impedance of cable lines that can be treated as infinitely long. This certainly does not mean that soil resistivity estimation based only on the geological structure of the local soil should not be careful as far as possible. Since in practical conditions soil resistivity along the cable line is normally not homogeneous by applying Eq. (2.54), it is necessary to adopt its equivalent value, which ensures final results slightly on the conservative side.

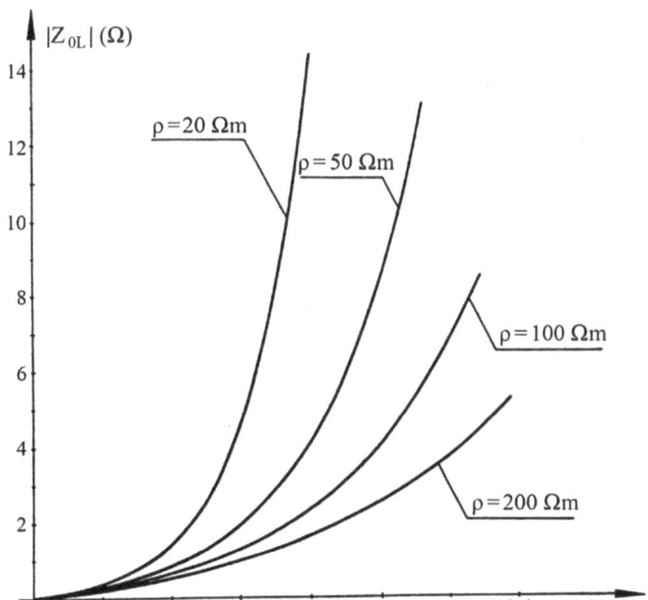

Figure 2.18 Transfer impedance of cable line metal sheath depending on *L* and for different values of *ρ*.

The developed analytical expressions here are used as a basis for the only developed analytical method for analysis of the complex grounding systems formed by uncoated metal-sheathed MV cable lines outgoing from an HV/MV distribution substation [23,24].

REFERENCES

[1] G. Metzger, J. Vabre, Transmission Line with Pulse Excitation, Academic Press, New York, London, 1969.

[2] F. Dawalibi, G. Niels, Measurements and computations of fault current distribution on overhead transmission lines, IEEE Transactions on Power Apparatus and Systems 103 (3) (1984) 533–560.

[3] B. Thapar, S. Madan, Current for design of grounding systems, IEEE Transactions on Power Apparatus and Systems 103 (9) (1984) 2633–2636.

[4] J. Endrenyi, Analysis of transmission tower potentials during ground faults, IEEE Transactions on Power Apparatus and Systems 86 (10) (1967) 1274–1283.

[5] D. Garrett, J. Myers, S. Patel, Determination of maximum substation grounding system fault current using graphical analysis, IEEE Transactions on Power Delivery 2 (3) (1987) 725–732.

[6] M. Tibensky, L. Perfecky, Methods for RMS symmetrical station ground potentials rise calculations for protection of telecommunications circuits entering power stations, IEEE Transactions on Power Apparatus and Systems 100 (12) (1981) 4785–4794.

[7] Lj.M. Popović, Dissipating of the partial ground fault current across the shield wires of transmission lines, Electronic Power Systems Research 11 (l) (1986) 25–37.

[8] L. Levey, Calculation of ground fault current using an equivalent circuit and a simplified ladder network, IEEE Transactions on Power Apparatus and Systems 110 (8) (1982) 2491–2497.

[9] S. Lambert, Minimum shield wire size – fault current considerations, IEEE Transactions Power Apparatus and Systems 102 (3) (1983) 572–578.

[10] S. Sobral, V. Costa, M. Campos, D. Mukhedkar, Dimensioning of nearby substations interconnected ground system, IEEE Transactions on Power Delivery 3 (4) (1988) 1605–1614.

[11] H. Goci, S. Sebo, Distribution of ground fault current along transmission lines – an improved algorithm, IEEE Transactions on Power Apparatus and Systems 104 (3) (1985) 663–669.

[12] Lj.M. Popović, General equations of the line represented by discrete parameters, Part I: – steady state, IEEE Transactions on Power Delivery 6 (l) (1991) 295–301.

[13] Lj.M. Popović, General equations of the line represented by discrete parameters; Part II: – resonant phenomena, IEEE Transactions on Power Delivery 6 (1) (1991) 302–308.

[14] Lj.M. Popović, Recognition of resonance in long radial transmission lines with static VAR compensation using travelling waves, ETEP European Transactions Electrical Power 15 (1) (2005) 31–41.

[15] Lj.M. Popović, Practical method for evaluating ground fault current distribution in station, towers and ground wire, IEEE Transactions on Power Delivery 13 (1) (1998) 123–129.

[16] Lj.M. Popović, Practical method for evaluating ground fault current distribution in station supplied by an unhomogeneous line, IEEE Transactions on Power Delivery 12 (2) (1997) 722–727.

[17] Lj.M. Popović, New method for calculation of series capacitance for transient analysis of windings, in: Proc. MELECON Conference, May 1998, pp. 1042–1046. Tel-Aviv, Israel.
[18] Lj.M. Popović, Active length of cable lines with uncoated metal sheaths as long grounding conductors, ETEP European Transactions on Electrical Power 22 (2) (2012) 206–215.
[19] R.J. Carson, Wave propagation in overhead wires with ground return, Bell System Technical Journal 5 (1926) 539–554.
[20] J.R. Carson, Ground return impedance: underground wire with earth return, Bell System Technical Journal 8 (1929) 94–98.
[21] R. Rudenberg, Grounding principles and practices I – Fundamental considerations of grounding currents, Electrical Engineering 64 (1945) 1–13.
[22] Int Std, ref CEI/IEC 60908-3, Ed 0.3 2009-3, Short-circuits Currents in Three Phase A.C. Systems – Part 3: Currents During Two Separate Simultaneous Line-to-earth Short Circuits and Partial Short – Circuits Following Through Earth.
[23] Lj.M. Popović, Practical method for the analysis of earthling systems with long external electrodes, IEE Proceedings C 140 (1993) 213–220.
[24] Lj.M. Popović, Comparative analysis of grounding systems formed by MV cable lines with either uninsulated or insulated metal sheath(s), Electric Power System Research 81 (2) (2011) 393–399.
[25] Lj.M. Popović, The effect of partial compensation of the unfavorable influence of increased soil resistivity in long earthling conductors, in: Proc. CIGRE Symposium, June 1985. Brussels.

FURTHER READING

[1] S. Mangione, A simple method for evaluating ground fault current transfer at the transition station of a combined overhead-cable line, IEEE Transactions on Power Delivery 23 (3) (2008) 1413–1418.

CHAPTER 3

Ground Fault Current Distribution

3.1 REDUCTION FACTOR OF FEEDING LINES

3.1.1 Introduction

During a ground fault in a high-voltage (HV) network, high currents and raised potentials appear at places where they normally do not exist. Under such conditions, an HV power line represents a very complex electrical circuit containing many conductively and inductively coupled elements. At the place of a ground fault the fault current leaves the phase conductor and returns toward its sources in the power network by using all available paths. As a result, a ground fault current, for a fault anywhere in an HV power network, divides into several fractions, separately returning to the sources in the power system. Two of them are of special importance for the designer of a future HV substation. One of them is dissipated into the surrounding earth through the grounding system of a supplied substation and returns through earth, while the other returns through the neutral conductor(s) of the feeding line. The first one produces all potentials and potential differences (touch and step voltages) relevant for estimation of the safety conditions in the grounding system of a supplied substation, while the other causes thermal stresses in the neutral conductor(s) of the feeding line. Thus a correct estimation of ground fault distribution is necessary for at a least two reasons: for the correct design of the grounding system of a future substation and for the correct selection of the neutral conductor(s) of its feeding line(s). However, prior to this all available return paths must be considered.

With the aim of simplifying and standardizing the procedure of determination of the mentioned fractions of the ground fault current, a special parameter of the power lines is introduced in the professional literature, including corresponding technical standards. This parameter is called the reduction factor of the feeding line and is defined as the ratio of the part of the ground fault current returning through the earth surrounding this line and the total ground fault current. By this definition, it

Practical Methods for Analysis and Design of HV Installation Grounding Systems
ISBN 978-0-12-814460-2
https://doi.org/10.1016/B978-0-12-814460-2.00003-9

is assumed that grounding impedances at both line ends, or in the supply and supplied substation, are negligible.

Under this assumption used for the definition of the reduction factor, a fraction of the ground fault current that appears in the neutral conductor(s) is a consequence solely of inductive coupling between the phase conductor carrying the ground fault current and neutral conductors (ground wires of overhead lines or metallic sheaths of cable lines). Since this component of the fault current returns to the power system only through the neutral conductor(s), it does not contribute to the creation of any potential in the grounding system of a supplied substation. Therefore it may be said that the feeding line reduction factor expresses only the influence of inductive coupling between the phase conductors and neutral conductors on reduction of the current dissipated through the grounding system of a supplied substation into the surrounding earth.

Based on the previously given definition of the reduction factor, a part of the ground fault current dissipated into the earth through the grounding system of the supplied substation is equal to the product of the reduction factor of the feeding line and the total ground fault current [1−6]. Thus it can be said that the correctly obtained value of the reduction factor is of primary importance for correct estimation of safety conditions in the grounding system of a supplied substation. If the value of this factor is higher than the actual one used, the result is a too conservative estimate of the achieved safety conditions and possibly excessive (undue) expenses for satisfying the strictly prescribed safety criteria. If the value is lower than the actual one used, this may result in an erroneous conclusion that the safety requirements are fulfilled, even though this is not the case. These are the reasons why it is of prime importance for designers of future substation to have at their disposal a method to determine the value of the reduction factor of a feeding line as accurately as possible. Also, with the aim of correctly selecting the neutral conductor(s), the users of future HV feeding lines, mainly power distribution companies, should have at their disposal a method for accurately determining their reduction factor.

If the method of determination of the reduction factor of a cable line is such that the obtained values are significantly above the actual one, the power company will order from the manufacturer cables that have a larger cross-section of metal sheath than those that are sufficient. Bearing in mind that lengths of medium-voltage (MV) and HV cables introduced each year into power distribution networks all over the world are enormous and permanently growing, this undue increase in metal sheath cross-section is certainly a reason of unnecessary increased total expenditure.

It is also important to know that in three-phase MV cable networks, where normally cross-bonding (metallic sheaths transposition) is not applicable, each increase in the cross-section of cable metal sheaths increases the losses of electric energy caused by circulating currents, e.g., Ref. [5]. Since the circulating currents exist under normal operating conditions, these increased losses exist permanently during the whole period of cable line utilization. Bearing in mind the enormous total lengths of MV cable lines in a large distribution network, it is clear that this type of expense also deserves our full attention. Also additional and undue losses in the cable sheaths cause additional and undue cable heating, which results in a decrease in the current-carrying capacity of the cable lines and certainly in the whole distribution network.

With the aim of solving the problem of determining ground fault current distribution, extensive and continuous research has been carried out over the last five decades. The first developed methods for solving this problem are related to the case where the feeding line is constructed as an overhead line, e.g., Refs. [7—10]. Then, the method based on computer analysis determines the critical fault position, which yields the maximum earth current in a generating substation grounding system [9]. Somewhat later the analytical method developed in Refs. [10,11] solved the same problem. Subsequently, papers referring to the problem in special cases, i.e., when a feeding line appears as a longitudinal combination of one cable and one overhead section, have been published, e.g., Refs. [12—15]. Also several methods have been developed for solving the problem in cases where, because of high local soil resistivity and/or a high level of short-circuit currents, special measurements (e.g., bare copper wire laid in the same trench as the cable feeding line, counterpoises, etc.) with the aim of reducing the part of the ground fault current dissipating into the surrounding earth are considered indispensable, e.g., Refs. [16,17]. Finally, the papers [5,18] present the method developed for determination of ground fault current distribution when a feeding cable line performs as three single-core cables. The method takes into account participation of all three metal sheaths in ground fault current distribution for a fault in a supplied substation(s) and at any other place along the feeding line.

The solution given in Ref. [5] served as a basis for the proposal of Serbia and Montenegro for the revision of corresponding IEC documents: the International Standard IEC 60909—3, Ed. 2 (2003—09) "Short-circuit calculation in three-phase AC system—Part 3: Currents during two separate simultaneous single phase line-to-earth short circuits and partial short-circuit currents following through earth" and Technical Report IEC 60909-2, Ed. 1 (1992–09) "Short-circuit current calculation in three-phase AC system—Part 2: Data of electrical equipment for short-circuit calculations."

Finally, paper [7] presents the only so far known method for determination of ground fault current distribution when HV substations are located in urban or suburban areas, or where many relevant data necessary for determination of this distribution are uncertain or completely unknown. The problem appears as a consequence of the fact that many urban metal installations are situated belowground and cannot be visually determined or verified. On the basis of on-site measurements, the developed method compensates for all deficiencies of relevant data regarding metal installations involved in the fluctuating magnet field appearing around and along a feeding HV line during a ground fault. The developed analytical procedure is based on the fact that two measurable quantities, currents in one phase conductor and in one neutral line conductor, cumulatively involve the inductive effects of all known and unknown surrounding metal installations. Once these currents have been determined by measurements, the problem of determining any relevant part of a ground fault current becomes solvable by using a relatively simple calculation procedure. The presented quantitative analysis indicates the significant benefits that can be obtained by taking into account the presence of surrounding metal installations.

3.1.2 Feeding Cable Line With Only One Metal Sheath

To obtaina better insight into ground fault current distribution and the relationship between relevant factors and parameters, let us start our consideration from a simple case that cannot be encountered in practice. This is where substation A supplies substation B through a single-phase feeding line consisting of only one single-core cable. Assuming further that the ground fault place is in substation B and that the neutral point in this substation is ungrounded, the ground fault current I_f is distributed as shown in Fig. 3.1.

The notation has the following meaning:

U—voltage source in substation A,

I_f—ground fault current,

I_{sh}—fault current through the cable sheath,

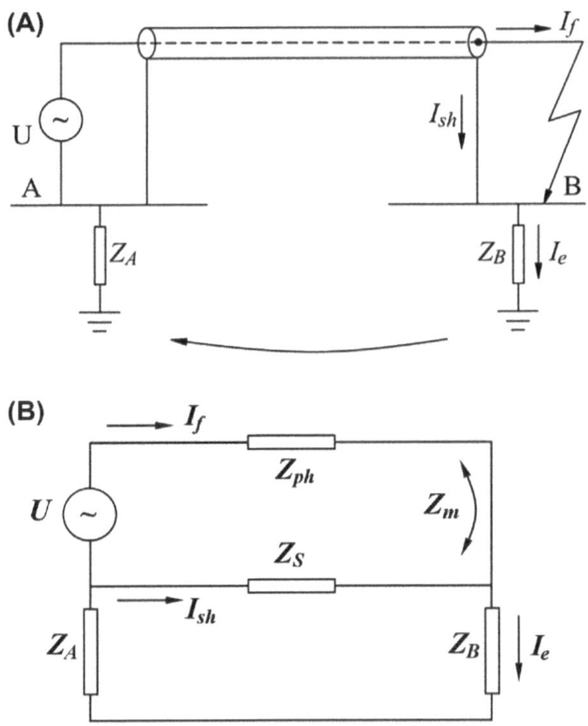

Figure 3.1 The simplest case of ground fault current distribution [5]. (A) Physical model of assumed cable line. (B) Equivalent circuit.

I_e—fault current returning through the earth,
Z_A (Z_B)—impedance of the grounding system of station A (B),
Z_{ph}—self-impedance of the phase conductor,
Z_S—self-impedance of the cable metal sheath, and
Z_m—mutual impedance between the cable metal sheath and the phase conductor.

On the basis of the equivalent circuit given in Fig. 3.1, one can see that the phase conductor and cable sheath with the earth as the common return path form mutually coupled circuits. Also, under the assumption that the grounding impedances at the line ends are negligible ($Z_A \approx Z_B \approx 0$), the fault current appearing in the line neutral conductor is a consequence solely of inductive coupling between this conductor and the phase conductor through which the ground fault current flows. Also it can be seen from the presented equivalent circuit that the current loops, cable conductor-earth and metal sheath-earth, are inductively coupled through impedance Z_m and conductively coupled through the grounding systems of substations A and B.

By using the equivalent circuit given in Fig. 3.1, one can write:

$$(Z_s + Z_A + Z_B)I_{sh} + (Z_m + Z_A + Z_B)I_f = 0. \tag{3.1}$$

By using this equation we can express the sheath current I_{sh} as a function of the total ground fault current, I_f. Also, for a better insight into the whole problem, we will, by using the decouple technique, represent this current through its two components, appearing as a consequence of the different and mutually independent influences, as in the following manner:

$$I_{sh} = I_i + I_c, \tag{3.2}$$

where:

$$I_i = -\frac{Z_m}{Z_S + Z_A + Z_B} I_{sh}, \tag{3.3}$$

$$I_c = -\frac{Z_A + Z_B}{Z_S + Z_A + Z_B} I_{sh}. \tag{3.4}$$

According to Eq. (3.3) it is obvious that I_i represents the component of the fault current appearing in the cable sheath only as a consequence of inductive coupling with the phase conductor (Z_m). The second component of the fault current I_c appears in the cable sheath only as a consequence of conductive coupling between the grounding electrode of substation A and the grounding electrode of station B.

On the basis of the presented equivalent circuit and using Eqs. (3.2–3.4) the distribution of the ground fault current can be expressed by the following coefficient:

$$r_t = \frac{I_e}{I_f} = \frac{Z_S - Z_m}{Z_S + Z_A + Z_B}. \tag{3.5}$$

This coefficient includes the effects of both inductive (Z_m) and conductive (Z_A and Z_B) coupling of the circuits presented in Fig. 3.1B. Since the influence of the conductive coupling depends on local factors such as Z_A and Z_B, the coefficient r_t cannot represent the parameter of the cable itself. To eliminate this influence we will assume, in accordance with the reduction factor definition, that impedances Z_A and Z_B are negligible ($Z_A \approx 0$ and $Z_B \approx 0$). Then, according to Eq. (3.5), we obtain that the reduction factor of the feeding line consisting of one single-core cable is given by:

$$r = \frac{Z'_S - Z'_m}{Z'_S}, \tag{3.6}$$

where impedances Z'_S and Z'_m are given per unit length. These impedances are, according to Ref. [4], determined by:

$$Z'_S = R'_{sh} + \omega\frac{\mu_0}{8} + j\omega\frac{\mu_0}{2\pi}\left(\frac{\mu_r}{4} + \ln\frac{\delta}{r_{sh}}\right),\ \Omega/\text{km} \tag{3.7}$$

$$Z'_m = \omega\frac{\mu_0}{8} + j\frac{\mu_0}{2\pi}\ln\frac{\delta}{r_{sh}},\ \Omega/\text{km}, \tag{3.8}$$

where R'_{sh} is the cable sheath resistance (Ω/km), r_{sh} is the medium radius of the cable sheath (m), ω is the angular frequency $\omega = 2\pi f$, and μ_0 is the magnetic permeability of vacuum, $4\pi\cdot10^{-7}$ (Vs/Am).

The equivalent earth penetration depth is determined by:

$$\delta = 658\sqrt{\frac{\rho}{f_n}},\ m, \tag{3.9}$$

where ρ is the equivalent soil resistivity along and around the cable line (Ωm) and f_n is the nominal power frequency.

Since the cable sheath is relatively thin, the reactance due to the flux inside the sheath can be disregarded. Thus we obtain for the reduction factor of the considered line the following approximation:

$$r \approx \frac{R'_S}{R'_S + \omega\frac{\mu_0}{8} + j\frac{\omega\mu_0}{2\pi}\left(\frac{\mu_r}{4} + \ln\frac{\delta}{r_{sh}}\right)}. \tag{3.10}$$

In accordance with the presented derivation procedure it is clear that this expression is quite correct for the cable lines performed by three-core cables. However, for many years it was used for lines consisting of three single-core cables as well. This practice was certainly not correct since it completely disregarded the participation of two cable line metal sheaths in the ground fault current distribution.

3.1.3 Constructive Characteristics of Lines Consisting of Single-Core Cables

The situation is somewhat different with electric power lines consisting of single-core cables. In a three-phase system a cable line consists of three such cables that can be laid in the ground in different ways, forming either a triangular or a flat formation. When they are in a flat formation, they may touch each other or may be laid at a certain separation from each other. Due to this the reduction factor of the cable lines consisting of three single-core cables depends not only on their designed characteristics, but also on their spatial formation in each concrete case.

Here, one should also mention the relevant constructive characteristics of single-core cables. As is well known, all polyethylene and cross-linked polyethylene (XLPE) cables for voltages above 1.8/3 kV should have a metallic screen or a metallic sheath. (For simplicity, later, only the second term will be used.) Its main function is to enclose the electric field inside the cable. This is accomplished by surrounding the insulation of a metal conductor with one or more metal tapes, or braid, or a concentric layer of wires, or a combination of wires and tape(s). A total cross-section of the sheath has to satisfy capacity requirements, especially those during unbalanced faults. In addition to this, cables planned for applications where they are exposed to large mechanical stresses also have mechanical protection, i.e., metallic armor.

The problem of determination of the reduction factor for cable lines consisting of three single-core cables was for a long time without an adequate solution. Manufacturers provide data regarding the reduction factor that refers only to one completely separated single-core cable or as produced and sold by manufacturers. However, distribution companies use them mainly in the formation of three single-core cables laid in the same trench, whereas their metal sheaths are usually mutually connected and grounded at the cable line ends. This obvious misunderstanding between manufacturers and electric power distribution companies certainly contributed to the problem that remained for a long time without an adequate solution. An additional reason is the fact that the correct value of the reduction factor of the cable line in general cannot be determined or verified only by measurements on site. This is due to the inductive influence of surrounding underground metal installations that are unavoidable in urban conditions.

For many years the analytical expression was used, which was adequate only for three-core cable lines or cable lines with only one metal sheath (Eq. 3.10). As a consequence, calculated values of the reduction factor were significantly larger than real ones. In many cases these values were almost 3 times larger than actual ones.

As an illustration, the ground fault current distribution in the case of a cable line consisting of three single-core cables is shown in Fig. 3.2.

The used notation has the following meaning:

A and *B*—supply and supplied substation,

I_f—ground fault current,

1—metal sheath of the cable carrying the ground fault current (later, the faulty cable),

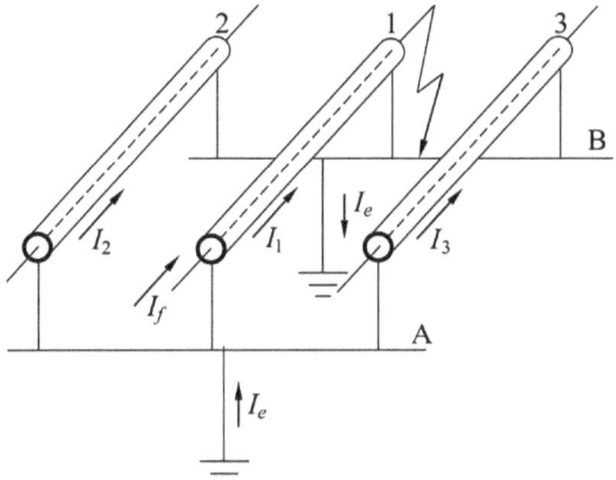

Figure 3.2 Ground fault current and its fractions [5].

2, 3—metal sheaths of the remaining two single-core cables,

I_1, I_2, and I_3—corresponding currents through the individual cable sheaths, and

I_e—part of the ground fault current I_f returning into substation *A* through the earth.

Thus it was necessary to derive analytical expressions enabling correct determination of the ground fault current distribution for a fault appearing in the supplied substation, as well as anywhere along the line. For this purpose a line made of single-core cables at ground fault conditions is presented as a complex electrical circuit containing conductively and inductively coupled conductors and with a common return path through the earth [5,6].

Under normal operating conditions, some currents also circulate in the line metal sheaths grounded at both ends. They appear because the magnetic fields produced by the load currents in each of the phase conductors do not mutually cancel completely, as is the case with magnetic fields produced by symmetrical phase currents in the metal sheath of three-core cables. These so-called circulating currents are larger if the spatial arrangement of the single-core cables is less symmetric and if the separations between the cables, because of better cooling, are larger. Certainly, they cause permanent losses in the cable sheaths and additional heating in all three cables. Also, as a harmful consequence of these currents, this additional cable heating reduces to a certain extent the current-carrying capacity of this type of cable lines.

The described problem can be significantly mitigated or even practically eliminated by applying the technical solution known as the cross-bonding technique. This technique is based on the transposition of the cable sheaths, or the phase conductors together with their metal sheaths, at certain places along the line. Since under practical conditions one cannot achieve, for instance, completely equal lengths of sections with transposed conductors (only metal sheaths or metal sheaths and phase conductors), cancellation of the circulating currents cannot be completed. Thus weak unbalanced currents circulate through the cable sheaths, earth, and grounding systems at the cable line ends. As a result, this solution is applied only when cable lines belong to an HV distribution network and mainly when the single-core cables are laid in a flat formation.

There is one more solution for the elimination of circulating currents. According to that solution the cable sheaths are grounded at only one point. However, application of this solution, because of the high and dangerous and harmful voltages that may be induced in the cable sheaths, is limited to very short lines, mostly those used for local connections, mainly within large generating stations.

3.1.4 Reduction Factor of a Line Consisting of Single-Core Cables

Let us consider this problem through an example of a cable line belonging to the so-called radial distribution network, whereby transmission substation A supplies HV distribution substation B. The neutral(s) in substation A is (are) either directly grounded or grounded through low impedance. Thus a ground fault in the feeding cable line or in a supplied substation acts as a short circuit and a large current may flow through the line phase conductor on the section between the fault point and grounded neutral point(s). Also let us assume that neutral point(s) on the HV distribution network to which the line belongs is (are) grounded only at substation A. Then, in the case of a ground fault in supplied substation B, parts of the fault current are presented by the equivalent circuit shown in Fig. 3.3.

The notation used has the following meaning:

F—fault place,

I_f—ground fault current,

I_1—fault current through the metal sheath of the faulty cable,

I_2 and I_3—fault currents through metallic sheaths of the other two cables,

I_e—fault current returning through earth,

Z_S—self-impedance of individual cable sheath,

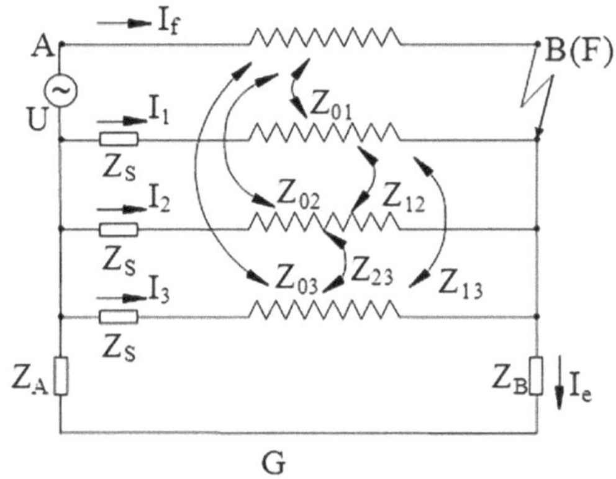

Figure 3.3 Currents during a ground fault in the supplied substation [2].

Z_{01}, Z_{02}, and Z_{03}—mutual impedances between the faulty phase conductor (denoted by index 0) and individual metallic sheaths (denoted by indexes 1, 2, and 3),

Z_{12}, Z_{13}, and Z_{23}—mutual impedances between individual metal sheaths,

Z_A (Z_B)—impedance of the grounding system of substation A (B), and

G—remote ground.

The balanced components of the ground fault current circulate only through the line phase conductors and their inductive influence on the nearby parallel circuits is so small that it can be disregarded. As a result, these currents and the sound phase conductors are disregarded in the presented equivalent circuit.

In accordance with the previously given definition of the reduction factor, in our further consideration we will assume that impedances Z_A and Z_B are negligibly small ($Z_A \approx 0$ and $Z_B \approx 0$). Also, for further consideration, it is necessary to mention that the current directions shown in Fig. 3.3 are arbitrarily taken.

By using the equivalent circuit in Fig. 3.3 and definition of the reduction factor, the following system of equations is obtained:

$$\begin{aligned}
&1.\ Z_S I_1 + Z_{12} I_2 Z_{13} I_3 + Z_{01} I_f = 0, \\
&2.\ Z_S I_2 + Z_{12} I_1 + Z_{23} I_3 + Z_{02} I_f = 0, \\
&3.\ Z_S I_3 + Z_{13} I_1 + Z_{23} I_2 + Z_{03} I_f = 0, \\
&4.\ r = \frac{I_e}{I_f} = \frac{I_f + I_1 + I_2 + I_3}{I_f}.
\end{aligned} \tag{3.11}$$

Since all of the line impedances are proportional to the cable line length, the first three equations in Eq. (3.11) can be divided by the length of the line. In this manner one obtains per unit length values of these impedances and eliminates the line length as superfluous data. On the basis of this observation we can treat all the impedances in Eq. (3.11) as the impedances per unit length and denote them with a prime ($'$). According to this, relative distribution of the ground fault current, I_f, does not depend on the value of this current, and the only unknown quantities are I_1, I_2, I_3, and r.

The most general case, regarding the values of mutual impedances, Z'_{02}, Z'_{03}, Z'_{12}, Z'_{13} and Z'_{23}, appears when the cables are laid in a flat formation with spacing (Fig. 3.4).

The notation used has the following meaning:

d—separation/distance between two adjacent cables.

Here it is assumed that the ground fault current is circulating through the core of one of the outer cables. This is the worst case from the standpoint of safety conditions in a supplied substation. Also the inductive influence on the surrounding metal installations is at its highest, because part of the ground fault current dissipating through the supplied substation grounding system takes its largest value.

The mutual impedance between two adjacent cables, denoted in the general case by Z'_m, is determined, according to, e.g., Ref. [4], by the following expression:

$$Z'_m = \omega\frac{\mu_0}{8} + j\frac{\mu_0}{2\pi}\ln\frac{\delta}{d}. \tag{3.12}$$

Mutual impedance between the two outer cables, denoted in the general case by Z'_{m2}, is also determined by expression (3.12), but, according to Fig. 3.4, "d" is replaced by "$2d$." Then, on the basis of the given analytical expressions and Fig. 3.4, it is clear that impedances Z'_{02}, Z'_{12}, and Z'_{23} are mutually equal $\left(Z'_{02} \approx Z'_{12} = Z'_{23}\right)$ and can be commonly denoted by Z'_m. Also it is clear that impedances Z'_{03} and Z'_{13} are approximately equal $\left(Z'_{03} \approx Z'_{13}\right)$, and that they can be commonly denoted by Z'_{m2}.

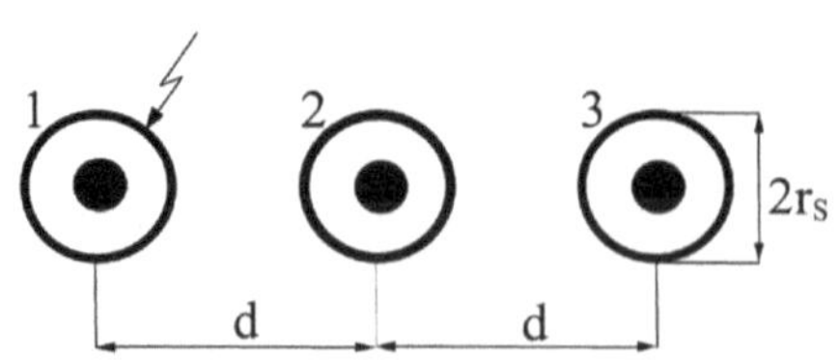

Figure 3.4 Single-core cable flat formation with spacing [2].

The system of Eqs. (3.11) can be solved by using the method of substitution of unknowns. Then, by using certain algebraic manipulations, one obtains the following expression:

$$r = \frac{R'_S}{Z'_S + Z'_m + Z'_{m2} - Z'_m \dfrac{Z'_m - Z'_{m2}}{Z'_S - Z'_m}}. \tag{3.13}$$

The obtained expression refers to the most general case when cables are laid in flat formation and on certain occasions separated from each other, and when a ground fault current passes through one of the outer cables. This case is critical from the standpoint of safety conditions because the part of the ground fault current dissipating through the grounding systems takes the largest value. On the basis of this expression, the value of the reduction factor in the general case depends on the self-impedance of the single-core cable sheath and mutual impedances between the single-core cables in the cable line.

Expression (3.13) can be easily modified for special cases, as it is a triangular formation that is frequently used in practice. In this case, cables are in a bundle and their positions are such that their centers are placed on top of an equilateral triangle (i.e., all three cables are positioned at a mutual distance: $d \approx 2r_S$). This means that in this case, impedance Z'_{m2} is equal to impedance $Z'_m\left(Z'_{m2} = Z'_m\right)$. In accordance with this, one obtains the following expression for the reduction factor of cables laid in triangular formation:

$$r = \frac{R'_S}{Z'_S + 2Z'_m}. \tag{3.14}$$

If by using Eqs. (3.7) and (3.12) Z'_s and Z'_m are substituted, Eq. (3.14) is transformed into:

$$r = \frac{R'_S}{R'_S + 3\omega\dfrac{\mu_0}{8} + j\omega\dfrac{\mu_0}{2\pi}\left(\ln\dfrac{\delta}{r_{sh}} + 2\ln\dfrac{\delta}{d}\right)}. \tag{3.15}$$

Finally, after certain algebraic manipulations, Eq. (3.15) is expressed in the following more concise form:

$$r = \frac{R'_S}{R'_S + 3\omega\dfrac{\mu_0}{8} + j3\omega\dfrac{\mu_0}{2\pi}\ln\dfrac{\delta}{\sqrt[3]{4}r_{sh}}}. \tag{3.16}$$

With a negligible error, expression (3.16) can also be used for cables laid in a flat formation. As will be seen from quantitative analysis (Section 3.4.1), the results obtained by Eq. (3.16) are slightly higher, i.e., slightly on the safe (desirable) side.

On the basis of expressions (3.10) and (3.16) it can be seen that the value of r will be larger if the value of the medium radius of the cable sheath, r_s, is larger, whereas according to Eq. (3.13) it can be seen that the value of r will be larger if the mutual distances between the single-core cables are larger.

One more effect, very desirable from the standpoint of grounding problems, should be pointed out here. When the natural conditions for solving grounding problems are less favorable, or when the specific soil resistivity ρ is higher, the value of r is lower, i.e., the part of the ground fault current passing through the earth I_e is smaller under other identical all other relevant conditions. According to expression (3.16), in the case of lines consisting of three single-core cables this effect is more pronounced than in the case of lines performed by three-core cable (Eq. 3.10).

On the basis of the given expressions it can be noticed that in the case of a flat formation of single-core cables, two different values can be obtained for the reduction factor of one cable line. One of these corresponds to the central faulted cable position, while the other corresponds to the other two faulty cable positions. The difference between these two values is larger if the mutual distances between the single-core cables are larger.

The mentioned two values are also obtained for a cable line with the cross-bonding technique applied. This is obvious on the basis of the circuit shown in Fig. 3.5; if we assume that the fault current I_f appears in the central single-core cable (denoted 2), then current I_1 and two currents

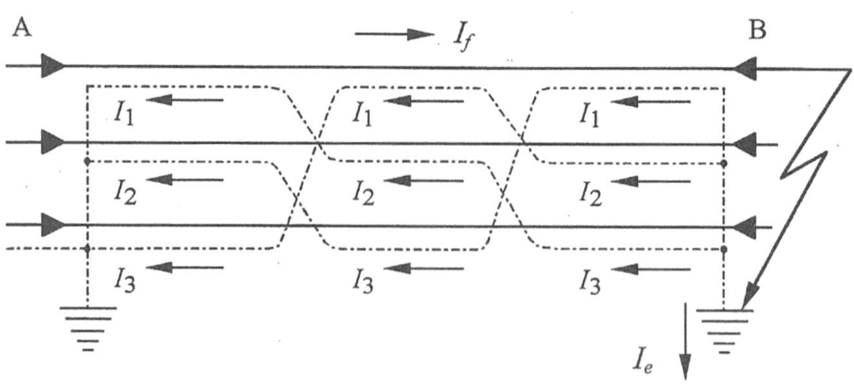

Figure 3.5 Cable line with applied cross-bonding [5].

I_2 will be induced in the cable sheaths instead of currents I_1, I_2, and I_3, as shown in Fig. 3.5.

The ground fault current distribution presented in Fig. 3.5 is also determined under the assumption that the grounding impedances at the line ends are equal to zero ($Z_A = Z_B = 0$).

Here also the following question can be interesting: where do the induced currents I_1, I_2, and I_3 flow if station A is a transition one and without grounded neutral point(s)? If one cable line supplies both substations (A and B) and has identical parameters along the whole of its length, according to Kirchhoff's rules these currents continue their path toward the supply substation through the grounding connections in substation A and through the cable sheaths of the rest of the common feeding line. Then, in situations where we have two or more transition substations and all other conditions are identical, the induced currents continue their path through the grounding connections of each of the transition substations and through the entire feeding line. However, the situation is different if any of the transition stations are supplied by a line having other constructive characteristics. For example, in the case of an overhead line, some of the induced currents continue their path toward their sources through the grounding system of this transition substation and the earth, while the rest of these currents continue their path through the ground wire(s) of the overhead section of the feeding line.

By applying the given analytical expressions it is possible to determine the reduction factor of a feeding line consisting of three single-core cables considering all three cable line metal sheaths. Here it is interesting to mention that the reduction factor of a feeding line represents a complex number, but for practical purposes its effective value is customarily used.

An analogous procedure can be used to develop analytical expressions for the determination of the reduction factor for specific cable lines consisting of two or more single-core cables per phase. On the basis of the procedure presented herein it is rather obvious that the value of the reduction factor in that case will successively decrease if we successively increase the number of the single-core cables per each phase conductor.

Under practical conditions the impedance of the substation grounding system is never equal to zero ($Z_A > 0$ and $Z_B > 0$). As a result, during ground faults, raised potentials appear at both ends of the feeding line. These potentials create a conductive component of the ground fault current in the cable sheaths and thus additionally reduce the fault current returning through the earth. This reduction mainly depends on local factors such as cable line length and grounding impedances at the feeding cable line ends. Consequently, this reduction cannot be treated as a parameter of the cable line itself and must be determined for each concrete case separately. In the case of a sufficiently long feeding line, the influence of the cable sheaths as a conductive connection with the grounding system of a nearby (supply or transition) substation can be neglected.

On the basis of the presented analytical procedure it is clear that the reduction factor of an HV or MV line consisting of three single-core cables is not technical data that can be obtained from a manufacturer of these cables. Its value depends, inter alia, on the formation/arrangement in which these cables will be laid. Thus only designers of future cable lines have all the data necessary for correct determination of their reduction factors.

In the time before than these expressions were derived, only the expression adequate only for the three-core cable lines was in application (Eq. 3.10). In this manner the influence of the currents induced in the remaining two cable sheaths on the value of the reduction factor of the feeding line is completely disregarded. Certainly, such estimation was excessively conservative and this led to unnecessarily high consumption of metals, either those used for cable sheaths or for the supplied substation's ground grids. As a final consequence, a significant quantity of useful and expensive metals (copper, aluminum, and lead) in refined states was needlessly returned to the earth.

However, all types of power cable lines, almost without exception, are being used in urban and/or suburban areas. In such environments, in addition to metal sheaths of cable lines, many other metal installations participate in the ground fault current returning to its sources in power system. Since they are situated mainly belowground their number and relevant technical characteristics are largely uncertain and partly or completely unknown.

3.2 REDUCTION FACTOR FOR LINES PASSING THROUGH URBAN AREAS

3.2.1 Basic Problem Description

In the case of a substation located in an urban area, correct determination of the ground fault current distribution is not possible on the basis of the formerly developed analytical expressions to determine the feeding line reduction factor. It is necessary to take into consideration the influence of the metal installations laid along the same street(s) as a feeding line and involved with the fluctuating and intensive magnetic field appearing along and around a feeding line during a ground fault in the supplied substation.

The permanently increasing sizes of modern urban distribution networks, causing higher operating and short-circuit currents, have been matched by overspreading networks of earth return circuits (different pipelines, neutral conductors of other electric power lines, etc.) situated in close vicinity to the HV and MV distribution lines. Spatial dispositions of all of these installations, which are determined mainly by dispositions of city streets, and small mutual separations result in intensive inductive and, in the vicinity of substations, conductive couplings of different unerground metal network types. The use of common routes (mainly street sidewalks) for various supply networks (electricity, water, gas, oil, telecommunications, etc.) unavoidably leads to the appearance of their mutual electromagnetic interaction that depends on many local factors. In urban conditions, metal installations of different basic functions, such as sheaths of different types of cable lines, neutral conductors of the LV networks, steel water pipes, reinforced concrete blocks of building foundations, etc., are situated in a relatively narrow space. Some of them are only in places in contact with the earth, whereas others are in effective and continuous contact with the earth. Also they are interconnected at many places, usually within each building, and their spatial dispositions are different in each particular case and vary along any concrete distribution line. Also most of them are laid under street sidewalks and much relevant data about them cannot be visually or otherwise determined or verified.

The grounding system of a distribution HV/MV substation consists of the substation grounding electrode and many outgoing MV cable lines acting as external grounding electrodes and/or conductive connections with the grounding systems of the supplied consumption (MV/LV) substations. Thus the grounding system spontaneously formed through outgoing MV cable lines involves a large urban area around an HV/MV distribution substation. Such a grounding system includes, through the terra-neutral grounding system in the LV network and consumer installations, many known and unknown metal installations typical for an urban area. As a consequence, the outgoing MV cable lines spontaneously become conductive connections with metal installations laid along the same street(s) as the HV feeding line. Due to their spatial disposition the surrounding metal installations are more or less inductively coupled with the neighboring feeding line and spontaneously participate in returning ground fault current toward the power system, or in reducing the fault current dissipated from the supplied substation grounding system into the surrounding earth. Thus the only task is: to determine this favorable effect at the design stage of an HV substation and at the moment of putting this substation into operation. However, the surrounding metal installations form a very complex network that is different in each particular case. Also this network has a different structure and configuration along the whole feeding line length. Thus it is clear that the actual physical model is very complex and involves numerous different metal installations over a large urban area involved by an HV distribution line.

In conditions where the ground fault current passes through one of the phases of a power line the portions of this current appear mutually in the surrounding metal installations due to their inductive connections with all conductors of the power line. The magnitude of these ground fault current portions depends on many factors such as their number, shape, dimensions, mutual disposition, and their disposition toward the considered power line conductors. Thus it is not difficult to imagine that the interaction in question is in reality very complex. It is complex to such an extent that it is impossible to analyze it purely by calculations, even under the assumption that all relevant data are known, which is certainly not the case in practice. However, the main difficulty appears because the mentioned installations are buried underground and therefore many relevant data remain uncertain, and in many cases even completely unknown. The same or similar can be said for soil composition along each HV line passing through urban and/or suburban areas.

Obviously, a complete and correct investigation of the influence of the surrounding metal installations requires an experimental analysis. However, the problem is in fact that besides the grounding function, the spontaneously formed grounding system in urban conditions acts also as a conductive connection with the metal installations surrounding the feeding line. As a result, during the ground fault an additional fraction of the ground fault current circulates in an axial direction through the grounding system elements. This fraction returns to the power system through the metal installations encompassed by the magnetic field produced by the ground fault current in the feeding line. The proximity of different grounding system elements in relation to the feeding line, as well as their interconnections within and around the HV/MV substation, produces different relationship between two ground fault current fractions in individual external grounding electrodes (metal sheath of the outgoing MV cables). Thus it is not difficult to imagine that the fraction induced by the magnetic field of the feeding line is distributed in a very complex way between the many external electrodes so that its immediate measurement is practically impossible. In addition, in practical conditions the current dissipating into the surrounding earth also cannot be correctly determined by measurements. The reason for this is the unavoidable influence of the nearby metallic installations on the accuracy of the results obtained by measurements. Thus it is also obvious that the considered problem cannot be solved only by field measurements. The problem was resolved by developing a method based on the results of field measurements and an additional analytical procedure, as will be shown here.

According to the former consideration, in the case of unbalanced operating conditions in an HV network, two separate current fractions circulate out of the considered electric power system. As an illustration, these two ground fault current fractions (I_e and I_i), separated by passing through the elements of the supplied substation grounding system, are presented in Fig. 3.6.

The used notation has the following meaning:

A(*B*)—supply (supplied) HV substation,

F—ground fault place (supplied substation),

I_f—ground fault current,

I_n—ground fault current fraction circulating through the line neutral conductor(s),

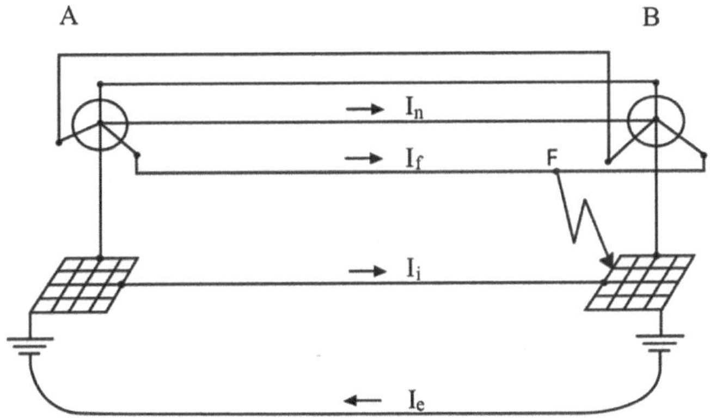

Figure 3.6 Main fractions of a ground fault current. *(Reproduced courtesy The Electromagnetic Academy: Lj.M. Popović, Ground fault current distribution when a ground fault occurs in HV substations located in an urban areas, Progress in Electromagnetic Research B 59 (2014) 167—179.)*

I_i—ground fault current fraction induced in the metal installations surrounding the feeding line, and

I_e—ground fault current fraction injected into the surrounding earth.

Both of the mentioned current fractions (I_i and I_e in Fig. 3.6) leave the electric power system through the elements of the grounding system of supplied substation *B*. Current I_e is dissipated into the earth, whereas current I_i continues its flow through the metal installations located along and around the feeding line. As a result, all potentially dangerous and harmful influences of the HV lines and supplied substation(s) upon their environment originate from these two ground fault current fractions.

Accordingly, determination of current I_e is of prime importance when estimating safety conditions within and in the vicinity of the supplied substation, while determination of current I_i is of prime importance when estimating inductive influence of a feeding line on the neighboring closely situated metal installations (pipelines, telecommunication lines, etc.). As is well known, more sensitive electronic systems face an increasingly hostile electromagnetic environment and electromagnetic coupling is an increasingly frequent cause of problems when it upsets data transmission; close attention needs to be paid to inductances of neighboring circuit loops. Since the process of division of these two ground fault current fractions occurs along many external grounding electrodes and belowground, none of these fractions can be separated and determined by direct or indirect measurements.

Each of the surrounding metal installations, together with earth as the common return path, forms one electrical circuit, while all of them together form a usually unknown number of conductively and inductively coupled electrical circuits. Thus the problem is: how could be the total induced current determined in these circuits? Since for all these circuits, on the basis of Kirchhoff's rules, one can write the corresponding system of equations with induced currents as unknown quantities, and since the analytical expressions necessary for the self- and mutual impedances of different types of underground conductors are known, it can be said that the formerly defined problem, in principle, was solved a long time ago. However, because of many practical difficulties, i.e., many uncertain and unknown relevant data necessary for determination the self- and mutual impedances of all attendant metal installations, this problem has for long time had no adequate solution, or a solution that would appreciate the favorable influence of surrounding metal installations. According to this, the problem can be also defined in the following manner: how can a method enabling compensation of the objective deficiency of a huge number of relevant data be found?

Finally, practical difficulty in estimating the influence of surrounding metal installations also stems from the fact that one basic piece of information, the value of soil resistivity of the surrounding area, cannot be determined exactly. Although there are several methods of measuring soil resistivity, not one of them is applicable in urban conditions. The reason for this is because surfaces of urban areas are already covered by buildings, streets, sidewalks, and many other permanently constructed objects, while belowground many known and unknown metal installations are already laid. The previously described problem is solved by developing a method based on the results of on-site measurements [2,3].

3.2.2 Equivalent Circuit of a Power Line Passing Through Urban Areas

On the basis of theoretical knowledge and performed experimental investigations, it was clear that each power line passing through urban and/ or suburban areas during a ground fault represents a more complex electrical circuit than can be assumed only on the basis of the total number of its conductors. The ground fault current circulating through one of the phase conductors induces in each of the surrounding metal installations one electromotive force. Thus a spontaneously formed electrical circuit can be imagined, in a general case, as a circuit with many mutually conductively and inductively coupled electrical circuits, with the earth as a common return path. With the aim of obtaining an insight into the completely formed, but usually not completely known, complex electrical circuit, let us assume initially that we know each of its constituent elements. Then, the total number of conductors that should be individually presented is equal to the total number of surrounding metal installations enlarged by the number of neutral line conductors and faulty phase conductor.

As an example, let us consider again the cable line consisting of three single-core cables, with all three metal sheaths grounded at both line ends. Also, let us assume that the total number of surrounding metal installations, including the neutral line conductor(s), is equal to an arbitrary number N. Then, the complete electrical circuit of the adopted line and its surrounding metal installations is presented as shown in Fig. 3.7.

The used notation has the following meaning:

U_a—auxiliary voltage source,

U_{n0}, U_{n1}, U_{n2}, ..., U_{nN}—voltages induced in an arbitrary (nth) circuit (metal installation) by the current in each of the surrounding circuits,

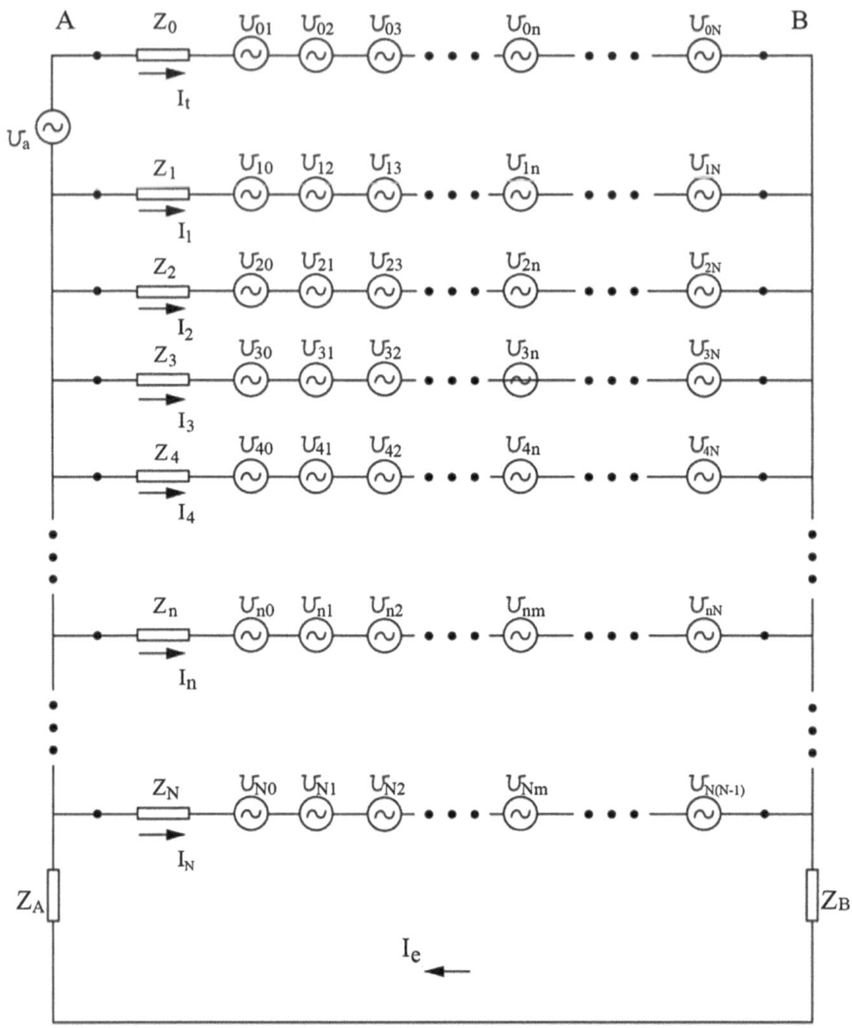

Figure 3.7 Complete equivalent circuit of the tested line. *(Reproduced courtesy The Electromagnetic Academy: Lj.M. Popović, Ground fault current distribution when a ground fault occurs in HV substations located in an urban areas, Progress in Electromagnetic Research B 59 (2014) 167–179.)*

I_t—simulated ground fault current through one of the phase conductors of the considered line,

I_1—current through the sheath of the cable carrying current I_t,

I_2, I_3—currents induced in the metal sheaths of the other two single-core cables,

I_4, I_5, …, I_n, …, I_N—currents induced in the individual surrounding metal installations,
Z_0—self-impedance of the phase conductor of the considered line,
Z_1, Z_2, Z_3—self-impedances of the cable metal sheaths,
Z_4, Z_5, Z_6, …, Z_N—self-impedances of the individual surrounding metal installations, and
N—total number of surrounding metal installations enlarged by the number of line neutral conductors.

As can be seen in Fig. 3.7, the presented equivalent circuit is composed of the self- and mutual impedances of the line conductors and all surrounding metal installations. The self-impedance of an arbitrary, nth, surrounding metal installation is, according to, e.g., Ref. [4], determined by:

$$Z'_n = R'_n + \omega_t \frac{\mu_0}{8} + j\omega_t \frac{\mu_0}{2\pi} \ln \frac{\delta_t}{r_n}, \ \Omega/\text{km}, \tag{3.17}$$

while mutual impedance between two arbitrary, nth and mth, surrounding metal installations is determined by:

$$Z'_{nm} = \omega_t \frac{\mu_0}{8} + j\omega_t \frac{\mu_0}{2\pi} \ln \frac{\delta_t}{D_{nm}}, \ \Omega/\text{km}; \quad m \neq n, \tag{3.18}$$

where R'_n is the resistance of an arbitrary, nth, surrounding metal installation, Ω/km, r_n is the radius of an arbitrary, nth, surrounding metal installation (m), D_{nm} is the distance/separation between two arbitrary, nth and mth, surrounding metal installations (m), ω_t is the angular test frequency, $2\pi f_t$, and δ_t is the equivalent earth penetration depth (m).

Current in an arbitrary surrounding installation, I_n (Fig. 3.7), induces in an also arbitrary (mth) surrounding installation a voltage, U_{mn}, which is, in general, determined by:

$$U_{mn} = Z_{nm} I_n, \quad m \neq n, \tag{3.19}$$

where Z_{nm} is the mutual impedance between the mth and nth surrounding metal installation.

As is known, HV distribution substations are mainly located in urban or suburban areas occupied by many underground metal installations acting as perfect grounding electrodes. Thus the relationships between values of impedances in the equivalent circuit of Fig. 3.7 are such that grounding impedances Z_A and Z_B can be disregarded ($Z_A \approx 0$ and $Z_B \approx 0$), which is in accordance with the formerly mentioned assumption for the line reduction factor definition (Section 3.1). On this basis it can be said that each current appearing in any of the line neutral conductors or any of the surrounding metal installations is a consequence solely of the mutual inductive coupling with each of all other surrounding conductors/installations. Also for further consideration it is necessary to say that the current directions shown in Fig. 3.7 are taken arbitrarily.

By using the complete equivalent circuit shown in Fig. 3.7 it is possible to write the system of *N*-independent equations. On this basis and considering the known values of the anticipated actual ground fault current I_F (available on the basis of a separate study) and the values of all self- and mutual impedances, it is possible determine each of the induced currents. However, it is sufficient for only one of the self- or mutual impedances in the presented complete equivalent circuit not to be known and the whole problem becomes unsolvable. Unfortunately, because of the previously mentioned practical difficulties, numerous data necessary for determining the self- and mutual impedances as well as the total number of surrounding metal installations in almost all practical cases cannot be determined and known. Thus with the aim of solving this problem a completely new approach was necessary.

On the basis of the completed equivalent circuit, it is not difficult to see that each individual current in the involved surrounding conductors (line neutral conductors and surrounding metal installations) appears as a consequence of the inductive influence of the test current through the line phase conductor I_t and the currents induced in all other surrounding conductors and installations. Thus each of the presented currents cumulatively involves the inductive influences of all other, known or unknown, currents presented in Fig. 3.7. Also it is clear that the value of each of the induced currents depends on all constitutive elements and parameters of the spontaneously formed electrical circuit, even those whose contribution is practically negligible. However, these facts although important are not sufficient for solving the problem or for determining the actual current circulating solely through the earth, I_e. They only indicated in which direction the desired solution could be found, or which quantity cumulatively involves all other relevant but unknown data and quantities.

The desired solution was achieved by measuring currents I_t and I_1 and by substituting all surrounding metal installations by only one equivalent conductor, imagined as a cylinder placed around and along the entire length of the considered line. Somewhat later, for the sake of simplifying the calculation procedure, it was shown that it was simpler when this fictitious conductor also involved metal sheaths of the two remaining single-core cables, through which the simulated ground fault (test) current did not circulate.

In the case of an overhead feeding line the number of neutral conductors is usually only one and, in accordance with this, the corresponding equivalent circuit is also reduced. However, the number of surrounding metal installations in this case is also unknown and as a result the total number of individual currents in the corresponding equivalent circuit can also be presented by an unknown number N.

3.2.3 Additional Neutral Conductor of an HV Cable Line

As has been previously shown, each power line passing through urban and/or suburban areas during an actual or simulated ground fault can be represented by the complete equivalent circuit shown in Fig. 3.7. In reality, metal installations presented by this circuit are mainly situated belowground or at places that usually are not accessible for performing any measurements. Thus many of the presented currents cannot be determined by direct measurements. Fortunately, this is not the case with currents passing through the power line conductors. These currents can be easily measured and this fact enables the collection of all necessary data for determining the overall influence of all surrounding metal installations, including metal sheaths of the two for the measurement purpose superfluous single-core cables on the current circulating solely through the earth, I_e. This can be done by simulating a ground fault in supplied substation B, and by measuring only two of the currents presented in the equivalent circuit of Fig. 3.7. These currents are the simulated ground fault current through one freely adopted single-core cable, I_t, and the current in its metal sheath, I_1. The necessary measurement circuit is presented in Fig. 3.8.

With the aim of simplifying the assumed complete equivalent circuit presented in Fig. 3.7, let us introduce an equivalent conductor substituting all surrounding metal installations, including metal sheaths of the two unused single-core cables. Then, the corresponding equivalent circuit of the considered cable line can be presented as shown in Fig. 3.9.

On the basis of the equivalent circuits presented in Figs. 3.7 and 3.9 it is clear that the number of unknown currents, which theoretically can be unlimited, is now reduced to only one current, I_c. Also the numerous known and unknown metal installations surrounding the adopted single-core cable is reduced to only one equivalent conductor. In this way all

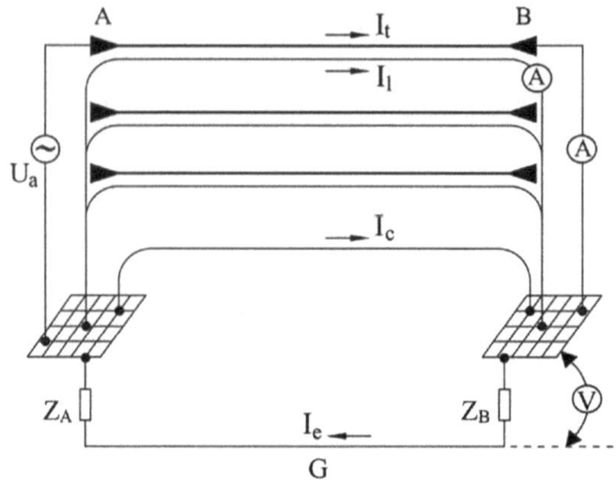

Figure 3.8 The principal circuit of the necessary measurements [3].

unknown but relevant data are reduced to the relevant data of only one equivalent conductor.

The relevant parameters of the introduced equivalent conductor can be determined under the condition that currents I_t and I_1 remain unchanged by substitution of the equivalent circuit of Fig. 3.7 with the equivalent circuit of Fig. 3.9.

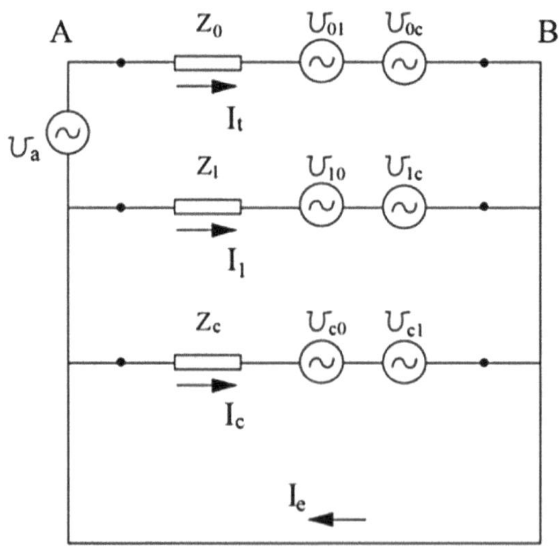

Figure 3.9 Equivalent circuit with only two neutral line conductors. *(Reproduced courtesy The Electromagnetic Academy: Lj.M. Popović, Ground fault current distribution when a ground fault occurs in HV substations located in an urban areas, Progress in Electromagnetic Research B 59 (2014) 167–179.)*

The used notation has the following meaning:

U_{1c} (U_{0c})—voltage that current I_1 (I_c) induces in the used phase conductor,

U_{10} (U_{1c})—voltage that current I_t (I_c) induces in the cable metal sheath,

U_{c0} (U_{c1})—voltage that current I_t (I_1) induces in the introduced equivalent conductor, and

I_e—current passing through the earth.

Also it is necessary that current I_c in Fig. 3.9 is equal to the sum of all currents in Fig. 3.7, with the exception of currents I_t, I_1, and I_e. Under these conditions, the demanded current I_e remains unchanged. Then, on the basis of the equivalent circuit presented in Fig. 3.9, the formerly defined conditions can be expressed by the following system of equations:

$$\begin{aligned} &1.\ Z'_{01}I_t + Z'_1 I_1 + Z'_{1C}I_C = 0 \\ &2.\ Z'_{0C}I_t + Z'_{1C}I_1 + Z'_C I_C = 0, \end{aligned} \tag{3.20}$$

where Z'_C is the self-impedance of the introduced equivalent conductor, per unit length, Z'_{0C} is the mutual impedance between the equivalent conductor and the phase conductor with the test current, per unit length, and Z'_{1C} is the mutual impedance between the equivalent conductor and the sheath of the cable carrying the test current, per unit length.

In the system of Eqs. (3.20), impedances Z'_1 and Z'_{01} are determined by the following expressions:

$$Z'_1 = R'_S + \omega_t\frac{\mu_0}{8} + j\omega_t\frac{\mu_0}{2\pi}\ln\frac{\delta_t}{r_{sh}},\ \Omega/\text{km} \tag{3.21}$$

$$Z'_{01} = \omega_t\frac{\mu_0}{8} + j\omega_t\frac{\mu_0}{2\pi}\ln\frac{\delta_t}{r_{sh}},\ \Omega/\text{km}. \tag{3.22}$$

With the aim of minimizing the number of unknown relevant parameters of the introduced equivalent conductor, the assumed physical appearance and space position of this conductor are imagined as a cylinder around and along the entire feeding line length. Under such assumptions, the considered cable line is modified into the simple model whose cross-section is shown in Fig. 3.10.

For the adopted appearance of the introduced equivalent conductor, the analytical expressions for impedances Z_C, Z_{0C}, and Z_{1C} are a priori known and determined by:

$$Z'_C = R'_C + \omega_t\frac{\mu_0}{8} + j\omega_t\frac{\mu_0}{2\pi}\ln\frac{\delta_t}{r_c},\ \Omega/\text{km} \tag{3.23}$$

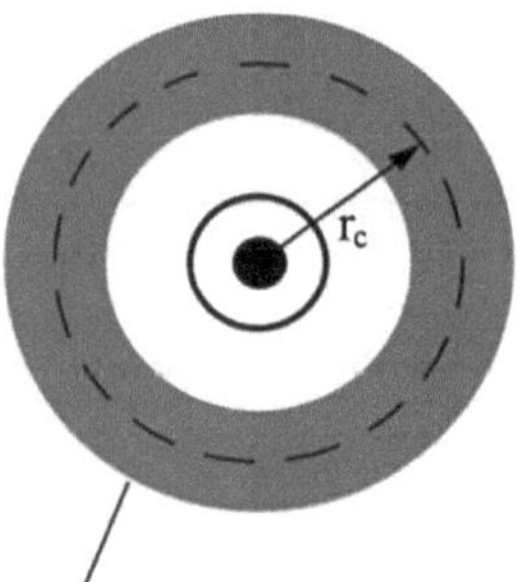

Figure 3.10 Cross-section of the imagined line model. *(Reproduced courtesy The Electromagnetic Academy: Lj.M. Popović, Ground fault current distribution when a ground fault occurs in HV substations located in an urban areas, Progress in Electromagnetic Research B 59 (2014) 167–179.)*

$$Z'_{0C} = Z'_{1C} = \omega_t \frac{\mu_0}{8} + j\omega_t \frac{\mu_0}{2\pi} \ln \frac{\delta_t}{r_c}, \ \Omega/\text{km}, \tag{3.24}$$

where R'_C is the resistance of the equivalent conductor (Ω/km) and r_c is the medium radius of the cylinder representing the equivalent conductor (m).

Since currents I_t and I_1 are obtained by measurements, they can be treated as known quantities, and the condition defined by Eq. (3.20) can be modified into the following:

$$\frac{Z_{0C}Z_{1C} - Z_C Z_{01}}{Z_C Z_1 - Z_{1C}^2} = \frac{I_1}{I_t}. \tag{3.25}$$

In this equation, the only unknown quantities, according to Eqs. (3.21–3.24), are R'_C and r_c. Since Eq. (3.25) gives the relationship between complex quantities, it can be presented as the following system of two equations:

$$\begin{aligned} \text{Re}\{(Z_{0C}Z_{1C} - Z_C Z_{01})I_t\} &= \text{Re}\{(Z_C Z_1 - Z_{1C}^2)I_1\}, \\ \text{Im}\{(Z_{0C}Z_{1C} - Z_C Z_{01})I_t\} &= \text{Im}\{(Z_C Z_1 - Z_{1C}^2)I_1\}. \end{aligned} \tag{3.26}$$

After solving this system of equations and determining the relevant parameters of the introduced equivalent conductor (R'_C and r_c), relations Eqs. (3.23) and (3.24) can be used for obtaining Z_C, Z_{0C}, and Z_{1C}, as well as currents I_c and I_e, by using the equivalent circuit of Fig. 3.9. Then, on the basis of this equivalent circuit and the reduction factor definition, the actual considered line reduction factor is determined by:

$$r = \frac{I_e}{I_t} = 1 - \frac{(Z_C - Z_{1C})Z_{01} + (Z_1 - Z_{1C})Z_{0C}}{Z_1 Z_C - Z_{1C}^2}, \tag{3.27}$$

or, according to Eqs. (3.21) and (3.22), in somewhat more compact form:

$$r = 1 - \frac{R'_C Z'_{01} + \left(Z'_1 - Z'_{1C}\right) Z'_{0C}}{Z'_1 Z'_C - Z'^2_{1C}}. \tag{3.28}$$

In the case of the line consisting of three single-core cables, the reactive parts of impedances Z_1 and Z_{1C} are approximately equal ($X_1 \approx X_{1C}$) and, instead of Eq. (3.28), the following approximation can be also used:

$$r \approx 1 - \frac{R'_C Z'_{01} + R'_1 Z'_{0C}}{Z'_1 Z'_C - Z'^2_{1C}}. \tag{3.29}$$

Finally, when the metal sheaths of two other single-core cables are disconnected (Figs. 3.7 and 3.8), under an unreal assumption that the cable sheaths do not exist ($Z_1 \rightarrow \infty$, $R_1 \rightarrow \infty$, $Z_{01} = 0$, and $Z_{1C} = 0$), the line reduction factor, existing only as a consequence of surrounding metal installations, is:

$$r_{mi} \approx \frac{Z'_C - Z'_{0C}}{Z'_C}, \tag{3.30}$$

or, in more direct form:

$$r_{mi} \approx \frac{R'_C}{R'_C + \omega_t \frac{\mu_0}{8} + j\frac{\omega_t \mu_0}{2\pi} \ln \frac{\delta_t}{r_C}}. \tag{3.31}$$

Here it is interesting to mention that thanks to the imagined physical appearance and space position of the introduced equivalent conductor, only two of its parameters are sufficient for allowing the problem solution to be possible. For any other freely imagined appearance and space position of this conductor, the considered problem could not be solvable because the number of the unknown quantities would be greater than the number of equations that can be written.

Finally, having in mind the assumptions ($Z_A \approx Z_B \approx 0$) made during derivation of the system of Eqs. (3.26), it is necessary to mention that the presented method is applicable only in cases where the following condition is satisfied:

$$Z_A \approx Z_B << Z'_S L, \tag{3.32}$$

where L is the length of the cable line between substations A and B.

In other words, the current appearing in the metal sheath as a consequence of only the difference of potentials of the grounding electrodes of the supply and supplied substations should be negligible. In practice, quantitative relationships between the relevant parameters are such that this criterion, Eq. (3.32), is satisfied for almost each substation and its feeding line in an HV cable network. Simultaneously, it means that in practical conditions, the influence of the feeding line length on the considered current distribution can be disregarded.

3.2.4 Additional Neutral Conductor of an Overhead Line

In contemporary HV distribution networks, overhead lines are also applied, especially in areas of a somewhat lower degree of urbanization. As a result, in the case of an overhead line passing through urban areas, data regarding the actual reduction factor of this line also cannot be obtained by using relatively simple analytical expressions, e.g., Ref. [4].

In this case, in comparison with the previous one, instead of the cable phase conductor and its metal sheath, we consider the phase conductor and ground wire(s) of an overhead line. Since the analogy is obvious, a similar equivalent circuit can be used as in the previous case. It is shown in Fig. 3.11.

The used notation has the following meaning:

U_{0w} (U_{0c})—voltage that current I_w (I_c) induces in the phase conductor carrying the test current,

U_{w0} (U_{wc})—voltage that current I_t (I_c) induces in the overhead line ground wire,

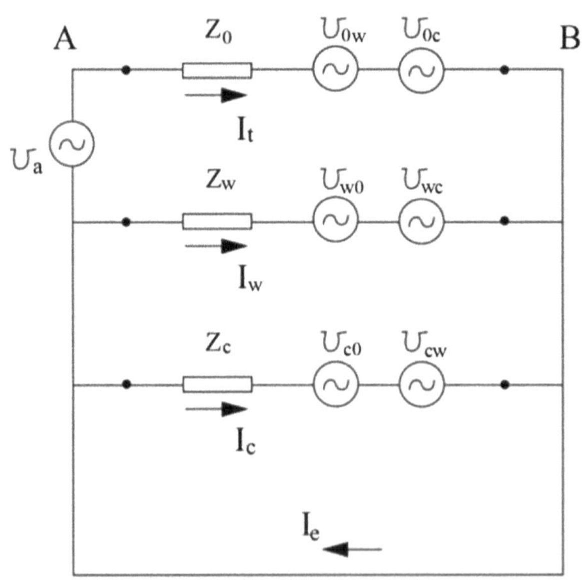

Figure 3.11 Equivalent circuit with the additional neutral conductor.

U_{c0} (U_{cw})—voltage that current I_t (I_w) induces in the introduced equivalent conductor, substituting all surrounding metal installations,
I_t—test current through one of the phase conductors,
I_w—current induced in the ground wire,
I_C—current induced in the additional neutral conductor, and
Z_w—self-impedance of the ground wire(s).

Self-impedance of the ground wire Z_w and mutual impedance between the ground wire(s) and the phase conductor Z_{0w} used for measurements are determined by:

$$Z'_W = R'_w + \omega_t \frac{\mu_0}{8} + j\omega_t \frac{\mu_0}{2}\left(\frac{\mu_r}{4} + \ln\frac{\delta_t}{r_w}\right),\ \Omega/\text{km} \tag{3.33}$$

$$Z'_{0W} = \omega_t \frac{\mu_0}{8} + j\omega_t \frac{\mu_0}{2}\left(\frac{\mu_r}{4} + \ln\frac{\delta_t}{d_w}\right),\ \Omega/\text{km}, \tag{3.34}$$

where R'_w is the resistance of the ground wire (Ω/km), r_w is the equivalent ground wire radius of one or more ground wires (m), d_w is the distance between the ground wire(s) and the phase conductor carrying the test current (m), ω_t is the angular test frequency, $2\pi f_t$, and μ_r is the relative permeability of the ground wire.

In this case it will be assumed that the introduced additional (equivalent) neutral conductor is also cylindrical in form and that its center is in the middle of the fictitious line connecting the ground wire(s) and the used phase conductor. On the basis of these assumptions, one can obtain the relevant parameters of the introduced equivalent neutral conductor by using the similar calculation procedure as in the previous case. However, it is important to emphasize that part of the test current appearing in the ground wire(s) as a consequence of the potentials of the grounding electrodes at the line ends can be significantly larger in this case. As a result, current I_W should be measured at the place that is sufficiently distant from substations A and B. This distance can be, according to Eq. (2.84), approximately estimated by:

$$N_a \approx \frac{3}{\ln\left(1 + \sqrt{\dfrac{|L_S Z'_W|}{R}}\right)}, \tag{3.35}$$

where N_a is the active length of the ground wire(s), expressed as the number of average line spans, R is the average tower footing resistance (Ω), and L_S is the length of one average span (km).

When a line is shorter than two active ground wire lengths, measurement of current I_w should be performed somewhere in the middle of such line.

Because of the significantly greater mutual distance between the phase conductors and their greater asymmetry toward the neutral conductor(s) and surrounding metal installations, in this case the critical phase conductor position cannot be known in advance (a priori). As a consequence it is necessary to repeat the whole previously presented procedure for each phase conductor position separately. The critical case, from the standpoint of safe conditions in the supplied substation, is certainly the phase conductor resulting in the largest value of current I_e.

3.2.5 Conductor Substituting All Surrounding Neutral Conductors

The considered, very favorable influence of the metal installations surrounding a feeding line laid in urban conditions is accompanied by an undesirable effect. During a ground fault in the supplied substation, too high a current can appear through one of the metal installations surrounding the feeding line on the whole of its length. Thus the following question can be posed: how can we determine the current induced in one of the surrounding metal installations seen separately?

Similar to the previously described procedure, this can be used to evaluate the current induced by an existing HV distribution line in the planned neighboring (e.g., in the same sidewalk or in the same street) insulated metal pipeline or in any of the existing surrounding installations. For this purpose, it is necessary to represent all surrounding metal installations, including all line neutral conductors, by only one equivalent neutral conductor, which is cylindrical in form and placed around the faulted phase conductor along the entire HV line length. On the basis of the equivalent circuit of Fig. 3.9, the relevant parameters of this equivalent conductor can be determined by:

$$R'_{eq} = \mathrm{Re}\{Z'_1 Z'_C - Z'^2_{1C}\} \tag{3.36}$$

$$r_{eq} = \frac{\delta}{e^{2\pi X'_{eq}/\omega\mu_0}}, \tag{3.37}$$

where R'_{eq} and X'_{eq} are real and imaginary parts of the self-impedance of the new equivalent conductor per unit length and r_{eq} is the radius of the equivalent neutral conductor in metres.

When the relevant parameters of the equivalent conductor are determined, the solution to the problem of the inductive influence on one of the

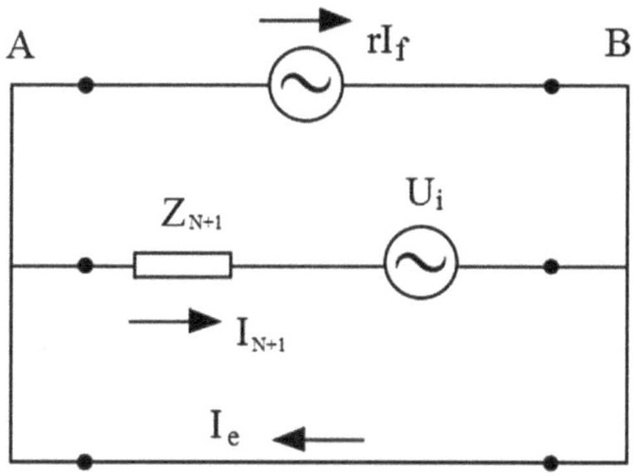

Figure 3.12 Inductive influence on one of the surrounding installations. *(Reproduced courtesy The Electromagnetic Academy: Lj.M. Popović, Ground fault current distribution when a ground fault occurs in HV substations located in an urban areas, Progress in Electromagnetic Research B 59 (2014) 167–179.)*

surrounding metal installations reduces to an extremely simple case. Bearing in mind the physical appearance and position of the introduced equivalent neutral conductor, the mutual impedance between this conductor and the planned nearby pipeline is the same as the mutual impedance between the faulted phase conductor and this pipeline. Thus the whole problem reduces to the simple case, which is represented by the equivalent circuit of Fig. 3.12.

The used notation has the following meaning:

rI_f—ideal current source,

I_{N+1}—current that will be induced in the planned pipeline,

$Z_{(N+1)}$—self-impedance of the planned pipeline, and

$Z_{(N+1)0}$—mutual impedance between the nearest (critical case) phase conductor and the planned pipeline.

On the basis of the equivalent circuit given in Fig. 3.12, the voltage that will be induced per unit length in the planned pipeline, U_i', can be determined by:

$$U_i' = -rZ_{(N+1)0}' I_f \tag{3.38}$$

When the relevant parameters of the equivalent neutral conductor are determined, the actual reduction factor can be determined by using, instead of Eq. (3.28), former expression (3.10), or:

$$r = \frac{R_{eq}'}{R_{eq}' + \omega\dfrac{\mu_0}{8} + j\dfrac{\omega\mu_0}{2\pi}\ln\dfrac{\delta}{r_{eq}}}. \tag{3.39}$$

Also mutual impedance between the nearest phase conductor and the planned pipeline, which will be laid along the same street(s) with an existing HV distribution line, is:

$$Z'_{(N+1)0} = \omega\frac{\mu_0}{8} + j\omega\frac{\mu_0}{2\pi}\ln\frac{\delta}{D} \tag{3.40}$$

where D is the foreseen separation/distance between the nearest phase conductor and the planned pipeline.

Since the ground fault current is a quantity known from a separate study, for determination of U_i' it is, according to Eq. (3.38), necessary to determine only the mutual impedance, $Z_{(N+1)0}$, and the actual reduction factor of the HV line. Since, by the using presented method, the currents induced in any of the surrounding metal installations can be determined, from a mathematical point of view the following is interesting to note: the presented method enables this problem to be solved, even in the case where the total number of relevant quantities (N) is unknown and, theoretically, unlimited.

On the basis of the presented analytical procedure it is obvious that the induced voltage and current will be smaller if the number of surrounding metal installations or a degree of urbanization of involved area is higher.

3.2.6 Application of the Presented Method

The previously presented method compensates for the objective deficiency of the necessary data regarding the surrounding metal installations through the field measurement of their cumulative effect upon the value of the test currents in phase conductors and in one of neutral conductors of the considered power line.

Certainly, application of the presented method assumes an already existing power line. However, data regarding the actual value of the reduction factor of an HV distribution line, or the ground fault current distribution in a supplied substation, are necessary at the design stage of this substation. This problem can be overcome by using the fact that, according to the equivalent circuit of Fig. 3.7, each of the neutral line conductors involves a cumulative inductive influence on all surrounding metal installations. This means that by taking into consideration this influence, a completely constructed line is not indispensable. Instead, it is sufficient only one suficiently long single-core cable. This would allow a provisory cable line to be laid on the surface of the soil along the foreseen path of the planned HV feeding line. The best approach for this purpose is to use an LV single-core cable (with a metal sheath), sufficiently flexible for different practically possible urban conditions and circumstances. Also, for simulating ground fault conditions and obtaining a corresponding electrical circuit with currents I_t and I_1 (Fig. 3.8), an auxiliary voltage source is necessary. In addition, for the purpose of realizing grounding electrode (building foundation) of the designed substation, if they do not already exist, one can use steel reinforcement in the foundation of one of the existing buildings nearest to the designed HV substation.

Since the surrounding metal installations are permanently attended and have an unchangeable spatial disposition, the electrical circuit formed by using the provisory cable line is practically the same as that presented by the equivalent circuit shown in Fig. 3.7. The only difference stems from the fact that in this case there is only one neutral conductor (cable sheath). By appreciating this difference, the voltage U_a, current I_t and all induced currents, according to the equivalent circuit of Fig. 3.7, are connected by the following equation:

$$U_a = Z_0 I_t + Z_{01} I_1 + \sum_{n=2}^{n=N} Z_{0n} I_n. \tag{3.41}$$

According to Eq. (3.41), the values of currents I_t and I_1 contain the cumulative effect of all, known and unknown, surrounding metal installations. Thus when the ratio between currents I_t and I_1 is obtained by the corresponding measurements, all surrounding metal installations can be substituted by only one equivalent neutral conductor. Then, the actual reduction factor of the considered HV line can be determined by treating this equivalent conductor as an additional neutral conductor and by using the calculation procedure analogous to the one presented in Section 3.1.4.

Since metal installations in urban areas are placed mainly under the soil surface, the determined influence of the surrounding metal installations will be slightly smaller in the case of a designed cable (underground) line or slightly greater in the case of a designed overhead line. Since the overhead line conductors are placed at a greater distance from the surrounding metal installations, the value of the radius of the equivalent cylindrical conductor representing the surrounding metal installations in the case of these lines should by enlarged by the average pole height. Because of this approximation the accuracy of the presented method is somewhat smaller at the design stage of future HV distribution lines.

Although there are several methods of measuring soil resistivity, none of them can give sufficiently trusted results when applied in urban conditions. The reason originates from the fact that surfaces of urban areas are already covered/occupied by buildings, streets, sidewalks, and many other permanently constructed objects, while belowground many known and unknown metal installations already exist. Thus we are forced to adopt an approximate value of the equivalent soil resistivity, based on the main geological characteristics of the involved area, and use it in the necessary calculations. Here, the favorable circumstance is that the self- and mutual impedances are, according to Eqs. (3.33) and (3.34), only slightly dependent on the equivalent soil resistivity and completely accurate data regarding this factor has no practical significance. It is sufficient to know that, within the framework of possible values of soil resistivity, the lowest one should be preferred because it gives final results that are slightly on the safe side.

Having in mind that the presented analytical procedure is developed on the basis of the results of the on-site measurements, the presented method takes into account the influence of all relevant factors and parameters, including those whose influence is very small. Thus accuracy of the presented method depends mainly on the accuracy of the measurements of the test current in one adopted phase conductor and current induced in the line neutral conductor, which, in principle, can be freely chosen if the considered line has more neutral conductors (Fig. 3.7). This means that the presented method gives a correct estimation for any, from the standpoint of complexity and the number of unknown relevant data, practical situation. In the case of an existing HV line and if mutual separations between the phase conductors are too great, the necessary measurement should be performed for each of the phase conductors with the aim of obtaining the largest (critical) value of the actual reduction factor.

Some inaccuracy can also appear as a consequence of the inductive influence of nearby power distribution lines. This influence can be efficiently avoided by using a test current of somewhat higher frequency that can easily be discriminated from the omnipresent power frequency. The introduced error is small and gives final results that are also slightly on the safe side.

Finally, the described method enables us to obtain the value of the reduction factor for any practically possible surroundings, if the ground fault occurs in the supplied substation itself. However, for estimation of the critical thermal stress that a cable sheath should withstand without any damage, the ground fault current distribution at any point along the feeding HV cable line has also to be determined.

3.3 GROUND FAULT CURRENT DISTRIBUTION FOR A FAULT AT ANY PLACE ALONG A CABLE LINE

3.3.1 Feeding Cable Line With Only One Metal Sheath

Generally, the thickness of the metal sheath must satisfy some contradictory requirements, i.e., it must be large enough to withstand thermal stress in ground fault conditions, but not too large to cause expensive cable production and too high values of harmful circulating currents, existing during normal operation conditions of this type of cable line. Therefore it is economically justified to look for an optimal size when choosing the cross-section of the metal sheath for cables in a particular distribution network. To reach this goal we must have, inter alia, a method, as accurate as possible, for calculating the ground fault current distribution for a fault anywhere along the cable line.

Let us assume the simplest case where station *A* supplies substation *B* through a single-phase cable line and that only station *A* has grounded neutral point(s). Then, if we assume that the fault appears at an arbitrary point *F* along this line the ground fault current distribution at the fault place can be presented as shown in Fig. 3.13.

The notation used has the following meaning:

I_{ai}—current induced in the cable sheath, and

I_{ac} (I_{bc})—current appearing as a consequence of the conductive connection with the grounding electrode in substation *A* (*B*).

According to the adopted assumptions the total fault current I_f comes only from station *A*. Thus the inductive influence exists only on the cable section from *A* to *F*. If we present this influence, according to the

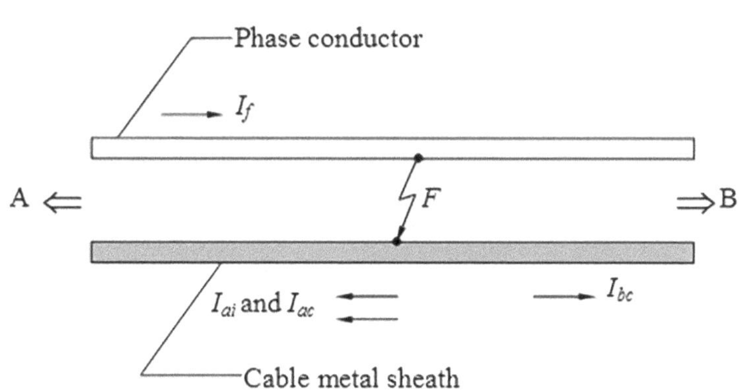

Figure 3.13 A fault in a cable line [5].

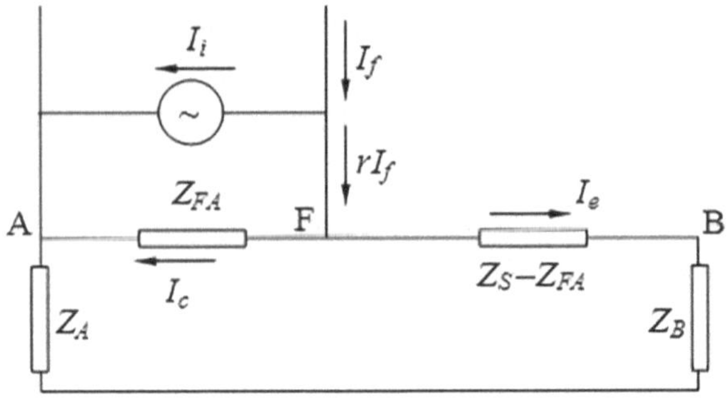

Figure 3.14 Equivalent circuit for a fault at any point along a cable line [5].

decoupling technique by an equivalent current source, determination of the ground fault current distribution becomes possible by using the equivalent circuit shown in Fig. 3.14.

The notation has the following meaning:

Z_S—self-impedance of the sheath for the whole line length,

Z_{FA}—self-impedance of the sheath from substation A to the fault place, F, and

r—reduction factor of the line for the fault at the end of the line.

We can represent Z_{FA} as a function of the cable section length by using the following expression:

$$Z_{FA} = lZ'_{sh}, \tag{3.42}$$

where l is the cable length from station A to F (km) and Z'_{sh} is the cable sheath self-impedance, per unit length (Ω/km).

On the basis of the equivalent circuit in Fig. 3.13 the remaining part of the fault current rI_f is divided into two fractions: currents I_c and I_e. Current I_c directly flows toward station A, while current I_e flows toward substation A through the grounding system of substation B and the earth. According to the given equivalent circuit and under the assumptions $Z_A \approx 0$ and $Z_B \approx 0$, we obtain:

$$I_e(L - l)Z'_{sh} = (rI_f - I_e)lZ'_{sh}, \tag{3.43}$$

where L is the total cable line length (km).

By using Eq. (3.43) the reduction factor for a fault at any point along the line is determined as:

$$r(l) = \frac{I_e}{I_f} = \frac{l}{L}r. \tag{3.44}$$

On the basis of the given expressions it can be seen that the reduction factor linearly increases with the distance of the fault place seen from station A. For a fault near to the beginning of the cable line, its value is close to zero and for a fault at the cable line end, substation B, it becomes equal to r (Eq. 3.10). Here it is important to say that for faults near the beginning of the cable line, almost all ground fault current passes through the cable sheath. As a result, the ground fault current at this place must be considered as critical from the standpoint of the terminal stress that a cable sheath has to withstand during the ground fault.

Now, let us consider the simplest case, from the standpoint of ground fault current distribution, of a three-phase cable line. This is a line constructed by one three-core cable. Since these cables have only a single metal sheath, it is clear that the derived analytical expression (3.10) refers to this type of cable line as well. Here it is important to note that inductive coupling between the sheath and the phase conductor(s) depends mainly on the constructive characteristics of the cables themselves and slightly on the local geological characteristics or equivalent soil resistivity, ρ. Thus in the case of this type of power cable line the reduction factor represents mainly a characteristic of the applied cable itself and can be accepted as manufacturer's data.

3.3.2 Feeding Cable Line Having Three Single-Core Cables

Let us assume that a fault appears at an arbitrary point F along a feeding cable line consisting of three single-core cables. For simplicity, we will assume that the ground fault current comes only from station A and that, in accordance with this, the induced component of the ground fault current in the metal sheath of the faulted cable appears only in the line section between substation A and the fault place. Also, for simplicity, let us neglect a relatively small current that can, in some practical cases, be injected into the ground at the fault place due to the damaged insulation sheath covering the cable metal sheath. Under such assumptions the ground fault current distribution in the faulted cable only and at the fault place is presented by the equivalent circuit shown in Fig. 3.13.

On the basis of Fig. 3.13 it is not difficult to imagine that in the case of a line having three single-core cables, the situation regarding ground fault current distribution becomes much more complicated. With the aim of obtaining a simple equivalent circuit as much as possible, let us apply the decoupling technique. In this way we can present separately the components of the fault current appearing as a consequence of the conductive connection with the grounding electrodes at the line ends. Also in this way we can present separately all components of the fault current appearing in the metal sheaths as a consequence of the inductive coupling with the faulted phase conductor. Starting from the previously adopted assumptions, on the basis of the decoupling technique, we obtain, for a fault at an arbitrary distance along the feeding line ($l = \overline{AF}$), the equivalent circuit as shown in Fig. 3.15.

The notation used has the following meaning:

U_a' $\left(U_b'\right)$—voltage source in the metallic sheath of the nearer sound cable as a consequence of inductive coupling between this metallic sheath and the faulted cable from A to F (from F to B),

U_a'' $\left(U_b''\right)$—voltage source in the metal sheath of the more distant sound cable as a consequence of inductive coupling between this metal sheath and the faulted cable from A to F (from F to B),

$(1 - r)I_f$—current source representing the fault current appearing in the metal sheaths as a consequence of their inductive coupling with the faulted phase conductor,

I_a and I_b—left and right current returning into the power system through the metal sheath, and appearing as a consequence of its conductive connections, and

F' and F''—points on the two sound cables nearest to the fault place, F.

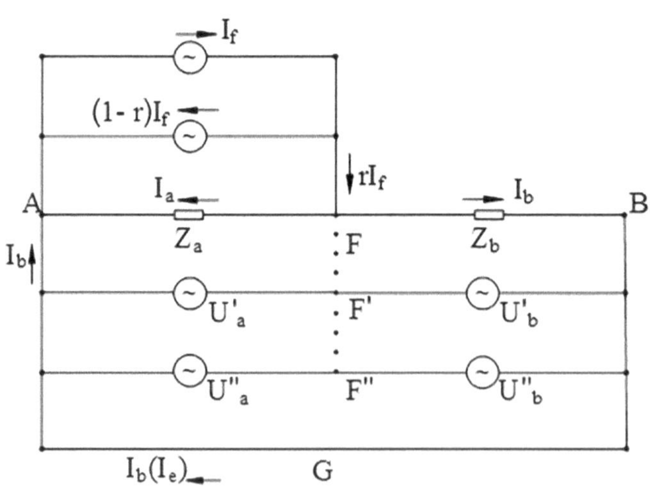

Figure 3.15 Equivalent circuit in the case of a fault at an arbitrary point along the line [6].

On the basis of the given equivalent circuit, the voltages induced in sections l and $(L - l)$ of the metal sheaths of the sound cables are mutually equal and determined by the following equations:

$$U'_a = U'_b = \frac{l(L-l)}{L} Z'_m r I_f, \tag{3.45}$$

$$U''_a = U''_b = \frac{l(L-l)}{L} Z'_{m2} r I_f. \tag{3.46}$$

Since, according to the given equivalent circuit, these voltages act in mutually opposite directions, they completely cancel each other. Thus, regarding currents in the presented equivalent circuit, we may consider that these voltage sources do not exist. At the same time, this means that current I_e is equal to I_b.

Then, for the remaining part of the circuit shown in Fig. 3.15, we can write the following system of equations:

$$\begin{aligned} &1.\ I_a + I_b = r(l) I_f, \\ &2.\ \frac{I_a}{I_b} = \frac{L-l}{l}. \end{aligned} \tag{3.47}$$

Finally, according to the definition of the reduction factor and this system of equations, one obtains:

$$r(l) = \frac{I_e}{I_f} = \frac{I_b}{I_f} = \frac{l}{L} r, \tag{3.48}$$

where $r(l)$ is the reduction factor for a fault at an arbitrary distance l along the line, seen from substation A.

As can be seen, expression (3.48) gives the relationship between the reduction factor when the fault place is at any point along the line and the reduction factor when the fault is at the line end, i.e., in the supplied substation. It shows that the reduction factor for the fault anywhere along the line is a linear function of the line length to the fault place. This observation represents a foundation for several conclusions important for current electrical engineering practice. They can be defined in the following manner:

- From the standpoint of safety conditions in the supplied substation, the critical faults are those within the substation itself.
- From the standpoint of inductive influence upon exposed metal installations, the critical faults are also those within the supplied station.

Then, since expressions (3.44) and (3.48) have the same mathematical form, another important conclusion should be pointed out:

- The function $r(l)$ does not depend on the number of neutral conductors.
  This means that $r(l)$ involves the influence of surrounding metal installations if r is determined by using the procedure described in Section 3.2.3. This is obvious on the basis of Fig. 3.15 as well.

Finally, on the basis of expression (3.48), it is not difficult to see that in the case of a ground fault near the supply station (at the beginning of the line) the reduction factor $r(l)$ reaches a value close to zero. This means that almost the whole fault current I_f returns to the power system solely through one of the cable line metal sheaths. As a result, this current and fault place must be treated as critical from the standpoint of the thermal stress that the cable sheath of the selected single-core cable has to withstand.

In practice, especially in the case of HV networks, the neutral point(s) in the supplied substation is (are) also directly grounded and as a result the ground fault current distribution along an HV cable line is more complicated. Namely, the total ground fault current represents the sum of two components: one arriving from station A (I_{fa}) and one arriving from station B (I_{fb}). In that case, according to the previous consideration and the principle of superposition, the total ground fault current is distributed in the manner shown in Fig. 3.16.

The notation used has the following meaning:

L—total cable line length,

l_a (l_b)—cable line section between the fault place and substation A (B),

I_{fa} (I_{fb})—left (right) component of the ground fault current I_f,

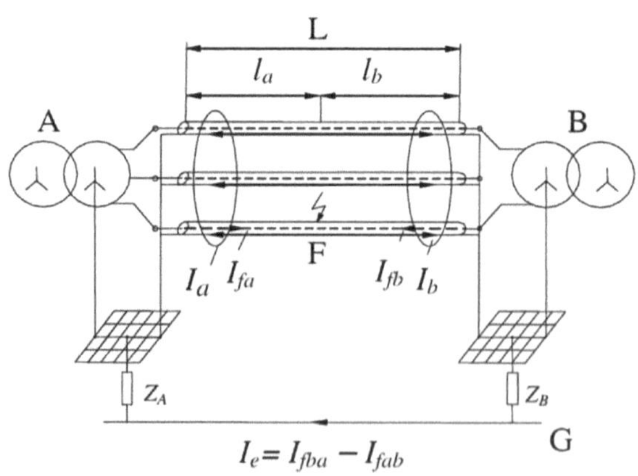

Figure 3.16 Ground fault at an arbitrary place along an HV cable line [6].

I_a (I_b)—left (right) fault current circulating through the metal sheath of the faulty cable, and

I_{fba} (I_{fba})—current as a consequence of the conductive connection between the fault place F and station A (B).

The ground fault current distribution shown in Figs. 3.15 and 3.16 is obtained under the assumption that the insulating sheath covering the metallic sheath at the fault place is not damaged. However, even if it is, in real conditions (large grounding resistance at the fault place) the current that would flow directly into the ground, as a consequence of such damage, would be negligible.

When the neutral point(s) in substation B is (are) grounded, one component of the fault current comes from station B. For the faults in substation B this component circulates only axially through the grounding system elements in station B and cannot create any potential in the grounding system of this substation. As a result, this ground fault current component independent of its value cannot be relevant from the standpoint of the safety conditions in supplied substation B. For faults along the cable, one part of this current returns to station B via the sheath, grounding system of substation A, and earth (Fig. 3.16). Thus it apparently reduces the value of the current passing via earth, i.e., the reduction factor for ground faults anywhere along the feeding cable line. However, this has no practical importance, because, as we have already seen, the critical fault position for safety conditions within and in the vicinity of a supplied station is this substation itself.

A similar consideration can be conducted for a more general and more complicated case, when at the side of station B we have an HV network with one or more grounded neutral points and/or a power station. In this case the ground fault current for the faults at any point along the cable line also consists of two components: one arriving from substation A and the other arriving from station B. However, the component from this substation now consists of two main subcomponents. One of them is passing through the grounded neutral point(s) in substation B and the other is returning into the power system through all of the other grounded neutral points in the HV network at the side of substation B. Certainly, this is similar to the component coming from supply substation A, which usually belongs to a large transmission network with many grounded neutral points.

Starting from the assumption that the influence of increased potentials in stations A and B is negligible ($Z_A \approx 0$, $Z_B \approx 0$), currents through the metal sheaths shown in Fig. 3.15 are determined by the following equations:

$$I_a = \left(1 - r\frac{l_a}{L}\right) I_{fa} + r\frac{l_a}{L} I_{fa} \frac{Z_F}{Z' l_a} + r\frac{l_b}{L} I_{fb} \frac{Z_F}{Z' l_b}, \tag{3.49}$$

$$I_b = \left(1 - r\frac{l_b}{L}\right) I_{fb} + r\frac{l_b}{L} I_{fb} \frac{Z_F}{Z' l_b} + r\frac{l_a}{L} I_{fa} \frac{Z_F}{Z' l_a}, \tag{3.50}$$

where Z_F is the grounding impedance of the sheath at the fault location, F.

This impedance is determined, according to the circuit shown in Fig. 3.14, by:

$$Z_F = \left(\frac{1}{Z'_{sh} l_a} + \frac{1}{Z'_{sh} l_b}\right)^{-1}. \tag{3.51}$$

Impedance Z_F represents a practically parallel connection of the two metal sheaths belonging to different line sections: one from the fault place to substation A and the other from the fault place to substation B.

As can be seen, currents through the sheath I_a and I_b consist of one induced and two components appearing as a consequence of the conductive connections between the fault place and the grounding electrodes at the line ends.

The currents induced in the faulted cable sheath are determined by:

$$I_{ai} = \left(1 - r\frac{l_a}{L}\right) I_{fa}, \tag{3.52}$$

$$I_{bi} = \left(1 - r\frac{l_b}{L}\right) I_{fb}. \tag{3.53}$$

Currents I_a and I_b have their maximum values if the faults are located very near to substation A ($l_a \to 0$) and substation B ($l_b \to 0$), respectively. By using the circuit in Fig. 3.16 it is not difficult to show that values of these currents at two critical locations (near station A and near station B) can be evaluated by the following approximations:

$$I_a(l_a \to 0) \approx I_{fa} + rI_{fb}, \tag{3.54}$$

$$I_b(l_b \to 0) \approx I_{fb} + rI_{fa}. \tag{3.55}$$

The larger of the currents obtained by these equations is to be applied for correct selection of the metal sheath capable of withstanding thermal stress during a ground fault. As impedances Z_A and Z_B are neglected, these equations give results that are slightly on the safe side. According to this, expressions (3.54) and (3.55) are necessary for correct dimensioning, i.e., selecting the metal sheath of the HV distribution cables, which is, as is well known, regulated through corresponding national standards.

The derived analytical expressions enable determination of the ground fault current distribution for a fault anywhere along the line composed of three single-core cables. By using them we can select an HV cable with a metal sheath that is in accordance with actual needs, i.e., large enough to withstand thermal stress in ground fault conditions, but not too large to cause expensive cable production and too high values of harmful circulating currents, existing during normal operation conditions of this type of cable line.

3.4 QUANTITATIVE ANALYSIS

3.4.1 Calculated Reduction Factor

By using the derived analytical expressions, let us perform a calculation of the reduction factor for a 110 kV cable line consisting of three single-core cables in a 50 Hz network. The relevant data on single-core XLPE-insulated cables with a copper sheath (copper screen of wires helically applied) are $r_S = 0.0455$ m and $R'_S = 0.18\ \Omega$/km (the sheath cross-section is 95 mm^2). The cables are laid in triangular formation ($d = 0.091$ m), whereas the other necessary data are $\delta = 931$ m or $\rho = 100\ \Omega$/m.

At first, let us consider values of the reduction factor obtained by Eq. (3.16) and data provided by the cable manufacturer for these cables before publishing [5,6]. Also let us consider the influence of the equivalent soil resistivity along the cable line in a wide range of possible values, as well as influence of different cable formations on the value of the reduction factor.

The results of the calculations of the reduction factor effective values are summarized and presented in Tables 3.1−3.4.

Table 3.1 Triangular formation by using Eq. (3.16)

ρ (Ωm)	30	100	300
$\lvert r\rvert$	0.1057	0.0992	0.0940

Table 3.2 Data provided by the manufacturer

ρ (Ωm)	30	100	300
$\lvert r\rvert$	0.2763	0.2618	0.2494

Table 3.3 Value of $|r|$ for different flat formations by using Eq. (3.13)

	ρ (Ωm)		
d(*m*)	**30**	**100**	**300**
0.091	0.1040	0.0988	0.0931
0.25	0.1114	0.1051	0.0951
0.50	0.1186	0.1107	0.1031

Table 3.4 Value of $|r|$ for different flat formations by using Eq. (3.16)

	ρ (Ωm)		
d(*m*)	**30**	**100**	**300**
0.091	0.1057	0.0992	0.0940
0.25	0.1141	0.1066	0.1005
0.50	0.1206	0.1122	0.1056

The values of $|r|$ obtained by taking into account only one metal sheath or provided by the cable manufacturer are almost three times larger, as shown in Table 3.2.

As can be seen, in all considered cases, values of the reduction factor increase as distance/separation between the cables increases and soil resistivity decreases.

Now, let us assume a substation that is at the design stage and that will be supplied by a 110 kV cable line with the required value of the reduction factor less than 0.16. The cable type will be XHE49-A; in the standard design its sheaths have an outer diameter of $d = 93$ mm and are made of copper strings with a total cross-section 3 × 95 mm^2. According to the manufacturer's catalog the reduction factor for this cross-section of the cable sheath is 0.297. By selecting the next larger cross-section offered in the catalog, 3 × 120 mm^2, the reduction factor, also according to the catalog, would be 0.235. Finally, the sheath cross-section of 3 × 185 mm^2 provides, also according to the catalog, the required values of the reduction factor equal to 0.157. If the length of a feeding 110 kV cable line per one 110/10 kV substation is approximately 4 km (an average value according to the data from the distribution network of Beograd, Serbia), the quantity of the copper necessary to decrease the reduction factor to the required level, below 0.16, is 9600 kg. However, according to the here presented analytical expressions the correct value of the reduction factor of the considered line consisting of the single-core cables with standard design of metal sheath (Cu 3 × 95 mm^2) has a value that is significantly lower, i.e., about 0.1. Here it should be mentioned that copper strings with a total cross-section of 95 mm^2 withstand fault currents of maximally 21.5 kA for a fault duration limited to 0.5 s. Thus, on the basis of the presented analytical expressions, application of the larger cable sheath cross-section offered by the manufacturer's catalog represents, from the standpoint of the required value of the reduction factor, nothing but redundant expenditure.

The current electric power industry standards mainly suggest a unique and semiempirical value for the reduction factor of different MV cable lines. This value is mainly significantly larger than the one that is obtained by Eq. (3.16), but the safety conditions of the supplied stations, because of urban surroundings, are always satisfied with a large reserve on the side of safety. Thus an analogous analysis can be performed for cables in MV networks. Sheath cross-sections of these cables are smaller (usually Cu 3 × 25 mm^2) and their reductions achieved by the analytical expressions derived here are smaller in comparison to cables in HV networks. However, the total length of the cables in a 10 kV network belonging to one 100/10 kV station is much larger (of the order of 120 km) and thus the total savings per a 100/10 kV station may be significantly larger.

3.4.2 Actual Reduction Factor

The grounding problem of HV/MV substations located in urban and suburban areas can appear in a very delicate form. As is well known, the reasons for this can be high local soil resistivity and/or a high level of short-circuit currents, vicinity of public objects, and a strictly defined and small yard that can be used for laying a grounding electrode. In such conditions it is of prime importance to take into account the favorable influences of the underground metal installations that exist in the urban surroundings independent of our need for solving this problem. Some of these installations are located around the HV feeding line and significantly improve the ground fault current distribution, while others are located around the supplied HV/MV substation and significantly reduce the impedance of its spontaneously formed grounding system. The previously presented method is the only one so far that allows designing, analysing, and testing of the grounding systems of HV/MV substations located in urban or suburban areas by taking into account the influence of the metal installation surrounding an HV feeding line.

Experimental investigations of the influence of metal installations typical for urban areas on the value of the feeding line reduction factor were performed on a cable line, which in series supplied two substations in the 110 kV distribution network of Beograd, Serbia. The length of the line to the nearer of the supplied substations measured from the supply substation was 2.32 km, while the total line length to the tested substation was 6.59 km. The line was made up of XHLP cables laid in a triangular formation on the whole line length. The cross-bounding technique for circulating current elimination was not applied to the line.

In the areas through which the line was passing the specific soil resistivity was estimated to be in a range from 30 to 50 Ωm. The line section between the supply substation and the transit one passed through an area with a lower degree of urbanization than the rest of the line. The phase conductors were made of aluminum with a cross-section of 1000 mm^2, while the metallic sheaths were made of copper strings with a total cross-section of 95 mm^2 and a medium diameter equal to 91 mm. The main elements of the grounding system in both of the supplied substations were the station building foundation and the 44 MV outgoing cable lines made up of cables with uncovered metal sheaths. The grounding impedances of the supply substation and both of the supplied substations were measured and had values between 0.02 and 0.03 Ω.

For the first time a test circuit was formed by using only a feeding line section between the supply and the transit substation. The length of this feeding line section was 2.32 km. The second test circuit was formed by using the whole considered feeding cable line. During the measurements in both of these cases, only the metal sheath of the cable with the test current was grounded at both of the test circuit ends.

By using the experimental measurement results and calculation procedure presented here, the actual reduction factor r_a obtained in the cases of the two experimentally investigated lines was:

- $r_a = -0.0267 - j0.0245$, or $|r_a| \approx 0.036$, for the shorter line (length: 2.32 km), and
- $r_a = -0.0225 - j0.0170$, or $|r_a| \approx 0.029$, for the longer line (length: 6590 m).

It can be seen that the effective values of the actual reduction factor were 65.9% and 72.6%, respectively, lower than the value obtained by using the corresponding analytical expression (3.16) through which it was not possible to take into account the influence of surrounding metal installations ($|r| \approx 0.105$).

If it is now assumed that the cables in the considered cases were laid in a flat formation, at a distance of $d = 0.5$ m, one obtains the following values:

- $r_a = -0.0275 - j0.0297$, or $|r_a| = 0.0405$ for the shorter line (length: 2.32 km) and
- $r_a = -0.0232 - j0.0202$, or $|r_a| = 0.0308$, for the longer line (length: 6590 m).

The actual values of the reduction factor in this case were lower by 78.0% and 84.2% than the value obtained by using the corresponding analytical expression (3.16), respectively.

In all considered cases the differences are still greater in comparison with the reduction factor values obtained only on the basis of the measurement results of currents induced in the cable metal sheaths, as was the usual but wrong practice before.

Thus, by ignoring the surrounding metal installations and using the existing analytical expressions for determination of the line reduction factor, as well as determining the reduction factor only on the basis of measurement results of currents induced in the cable sheaths, it is obvious that the values are excessively conservative. Also it is important to stress that the effects of reduction of the part of ground fault current passing through the earth are greater if the involved urban area is on a greater degree of urbanization.

On the basis of the presented method for taking into consideration the influence of the metal installations involved with a fluctuating magnetic field around and along the feeding line, and on the basis of the results of experimental investigation, the part of the ground fault current passing through the earth in a typically urban area is 3–6 times smaller than was previously considered. Accordingly, the actual values of the grounding system relevant parameters (touch and step voltages, as well as transferred potentials) are also smaller in the same proportion. This fact throws completely new light on the entire grounding problem of HV distribution substations and enables us to see it in a realistic framework. Thus, it can be said, that actual grounding problems of HV substations located in urban or suburban areas are much smaller than it was considered before.

Having in mind the similarity of urban conditions throughout the world and that the experimental investigations were performed in an area that can be considered as average from the standpoint of the achieved degree of urbanization, the former observations can be adopted as generally valid for the grounding systems of HV substations located in urban areas.

3.4.3 Critical Currents Through the Sheaths of a Cable Line

The diagram representing relative values of the currents induced in the cable sheath by currents $i_A(I_A/I_{fA})$ and $i_B(I_B/I_{fB})$ as a function of relative distance from station A (l_a/L) in the case of cable line triangular formation and ρ equals 100 Ωm is presented in Fig. 3.17.

As can be seen, the full lines representing i_A and i_B are symmetrical with respect to the middle of cable line. Also the dashed lines, representing the reduction factors r_A and r_B for the right and left ground fault current components, are symmetrical in relation to the middle of cable line.

In accordance with the given diagram, it is not difficult to conclude that current I_A flowing toward substation A reaches its maximum when the fault place is approaching the beginning of the line, while current I_B flowing toward substation B reaches its maximum when the fault place is approaching the end of the line. Here it should be mentioned that the value of current I_{fA} hyperbolically decreases with increasing fault distance from substation A, while current I_{fB} hyperbolically decreases with increasing fault distance from substation B.

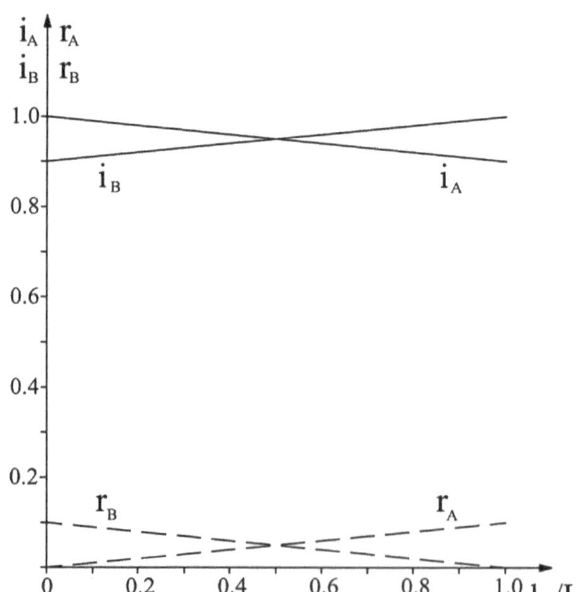

Figure 3.17 Relative values of the currents in a faulty cable sheath and the corresponding reduction factors [20].

3.4.4 General Observations

Single-core cables have been utilized in distribution networks of medium (10−35 kV) and high (66−132 kV) voltages all over the world for many decades. Although this is quite a long period of time, the problem of reduction factor determination for the feeding line consisting of three single-core cables was without adequate solution. After developing the analytical expressions presented here, it was possible to take into account the inductive influence of all metal sheaths on the ground fault current distribution for faults anywhere along power lines consisting of three single-core cables.

In accordance with previous considerations, estimation of the safety conditions in the supplied substations and in their vicinity was in the past excessively conservative. The estimated values of all potential differences on the grounding systems of the supplied substations were nearly 3 times higher than in reality. Depending on other relevant factors such as the level of ground fault currents, duration of cleaning time, and soil resistivity in each concrete case, this also leads to variable unnecessarily additional consumption of useful metals used for the cable sheaths and/or for the grounding electrodes of supplied substations. Significant amounts of copper, aluminum, and lead were returned in a refined state into the earth. In addition, in three-phase MV networks, where cross-bonding is not applicable for the elimination of the circulating currents, each unnecessary increase in the cross-section of the metallic sheaths increases energy losses caused by these currents. Since the circulating currents appear under normal operating conditions, these increased losses grow with arithmetical progression from year to year (each year numerous new MV cable lines are introduced into distribution networks all over the world). Also these additional losses cause additional heating of the cables, which results in a certain reduction of current-carrying capacity of the cable lines and of the whole distribution network. From this point of view, critical areas are certainly in the vicinity of HV/MV substations, where a large number of MV cables are concentrated in a narrow space, i.e., in a small number common trenches.

Since the use of single-core cables is increasing, especially cables with XLPE, it is realistic to expect that a significant saving of useful metals in the production of this type of cable will be accomplished by wide application of the here presented analytical expressions. Achieved savings will be increasingly larger with an increase in the total length of the applied cables, i.e., with the development of MV and HV distribution networks, as well as with permanently growing needs for the replacement of aged cables.

Bearing in mind all these facts, IEC—TC 73 (Technical Committee for "Short-circuit currents") accepted the proposal of Serbia and Montenegro for revision of the corresponding IEC documents International Standard IEC 60909—3, Ed. 2 (2003—09) and Technical Report IEC 60909-2, Ed. 1 (1992—09). By treating the presented analytical expressions as a new technology, IEC—TC 73 (at the meeting in Dresden, Germany, September 2004) decided to make a revision to the standard IEC—60909—3 (2003) just 1 year after the beginning of its implementation. The presented analytical expressions have been a foundation for the current version of the mentioned technical standard IEC 60909-3, Edition 3.0, 2009—02, "Short-circuit currents in three-phase AC systems—Part 3: Currents during two separate simultaneous line-to-earth short circuits and partial short-circuit currents flowing through earth" [4]. Through this technical standard, the presented analytical expressions become available for application in current electrical engineering practice worldwide. However, the considered problem has definitely not been solved in this way.

The method described in Section 3.2 determines the actual ground fault current portion dissipated through the grounding system of HV substations located in urban or suburban areas. Also the presented method enables determination of ground fault current distribution between feeding line neutral conductor(s) and the surrounding metal installations as a ground fault current return paths.

The reduced value of the feeding line reduction factor simultaneously means reduction by the same ratio of all potentials and potential differences (step and touch voltages) appearing on the grounding systems of the supplied HV substations. Thus one can conclude that the results of the here presented quantitative analysis throw a completely new light on the grounding problem of HV substations located in urban surroundings and dramatically change our former perception of its magnitude. Also, having in mind the similarity of urban conditions throughout the world and that the experimental investigations were performed in areas that can be treated as average/ordinary from the standpoint of the achieved degree of urbanization, this conclusion can be treated as generally valid for the safety conditions of distribution substations located in urban areas.

The only inconvenience arises from the fact that the actual ground fault current distribution for each line is dependent on local factors, i.e., specific characteristics and disposition of the metal installations in the area through which the feeding line passes. This means that to obtain an actual feeding line reduction factor, each distribution line passing through urban and/or suburban areas should be considered individually.

Since the cable lines are, almost without exception, applied in urban areas and since the effects of the surrounding metal installations are considerable, the presented method could be served as a foundation for revision of the current version of the corresponding technical standards [4].

REFERENCES

[1] Lj.M. Popović, Comparative analysis of grounding systems formed by MV cable lines with either uninsulated or insulated metal sheath(s), Electric Power System Research 81 (2) (2011) 393–399.

[2] Lj.M. Popović, Influence of the metallic installations surrounding the feeding cable line on the ground fault current distribution in supplied substations, IEEE Transactions on Power Delivery 23 (4) (2008) 2583–2590.

[3] Lj.M. Popović, Testing and evaluating grounding systems for substations located in urban areas, IET Generation Transmission and Distribution 5 (2) (2011) 231–238.

[4] Short-Circuits Currents in Three-Phase A.C. System-part 3: Currents during Two Separate Simultaneous Line-to-earth Short Circuit and Partial Short-C.

[5] Lj.M. Popović, Determination of the reduction factor for feeding cable lines consisting of three single-core cables, IEEE Transactions on Power Delivery 18 (3) (2003) 736–743.

[6] Lj.M. Popović, Improved analytical expressions for the determination of the reduction factor of the feeding line consisting of tree single-core cables, European Transactions on Electrical Power 18 (2) (2008) 190–203.

[7] Lj.M. Popović, Determination of the actual reduction factor of HV and MV cable line passing through urban and suburban area, FACTA UNIVERSITATIS-Series: Electronics and Energetics 27 (1) (2014) 25–39.

[8] J. Endrenyi, Analysis of transmission tower potentials during ground faults, IEEE Transactions on Power Apparatus and Systems 86 (10) (1967) 1274–1283.

[9] F. Dawalibi, G. Niles, Measurements and computations of fault current distribution on overhead transmission lines, IEEE Transactions on Power Apparatus and Systems 103 (3) (1984) 553–560.

[10] S.A.P. Meliopoulos, R. Web, E. Joy, S. Patel, Computation of maximum earth current in substation switchyards, IEEE Transactions on Power Apparatus and Systems 102 (9) (1983) 3131–3139.

[11] Lj.M. Popović, Practical method for evaluating ground fault current distribution in station, towers and ground wire, IEEE Transactions on Power Delivery 13 (1) (1998) 123–129.

[12] Lj.M. Popović, A practical method for evaluation of ground fault current distribution on double circuit parallel lines, IEEE Transactions on Power Delivery 15 (1) (2000) 108–113.

[13] S.T. Sobral, et al., Grounding potential rise characteristics of urban step-down substations fed by power cables – a practical example, IEEE Transactions on Power Delivery 3 (2) (1988) 1564–1572.

[14] S. Mangione, A simple method for evaluating ground-fault current transfer at the transient station of a combined overhead cable line, IEEE Transactions on Power Delivery 23 (3) (2008) 1413–1418.

[15] Lj.M. Popović, Practical method for evaluating ground fault current distribution in station supplied by an unhomogeneous line, IEEE Transactions on Power Delivery 12 (2) (1997) 722–727.
[16] S.T. Sobral, V.S. Costa, M.S. Campas, D. Mudhedkar, Dimensioning of nearby substations interconnected grounding systems, IEEE Transactions on Power Delivery 5 (3) (1988) 1605–1614.
[17] Lj.M. Popović, Efficient reduction of fault current through the grounding grid of substation supplied by cable line, IEEE Transactions on Power Delivery 15 (2) (2000) 556–561.
[18] A. Campuccia, G. Zizzo, A study of the use bare buried conductors in an extended interconnected earthing system inside a MV network, in: Proc. CIRED Conference, Turin, 2005.
[19] Lj.M. Popović, Ground fault current distribution when a ground fault occurs in HV substations located in an urban areas, Progress in Electromagnetic Research B 59 (2014) 167–179.
[20] Lj.M. Popović, Ground fault current distribution for the fault anywhere along the feeding line consisting of three single-core cables, in: Sixth in. Symposium, Nikola Tesla, Belgrade, Serbia, October 2006.

FURTHER READING

[1] R.J. Carson, Wave propagation in overhead wires with ground return, Bell Systems Technical Journal 5 (1926) 539–554.
[2] J.R. Carson, Ground return impedance: underground wire with earth return, Bell System Technical Journal 8 (1929) 94–98.
[3] R. Rudenberg, Grounding principles and practices I - Fundamental considerations of grounding currents, Electrical Engineering 64 (1945) 1–13.
[4] Lj.M. Popović, Active length of cable lines with uncoated metal sheaths as long grounding conductors, European Transaction on Electrical Power 22 (2) (2012) 206–215.
[5] IEEE Guide for Safety in Substations Grounding, Std. 80–2000.
[6] Lj.M. Popović, Practical method for the analysis of earthling systems with long external electrodes, IEE Proceedings C 140 (3) (1993) 213–220.
[7] Lj.M. Popović, Transfer characteristics of electric-power lines passing through urban and suburban areas, International Journal of Electrical Power & Energy Systems 56 (2014) 151–158.

CHAPTER 4

Different Cases of Nonhomogeneous Feeding Lines

4.1 LINE CONSISTING OF CONSTRUCTIVELY DIFFERENT SECTIONS

4.1.1 Introduction

During a phase-to-ground fault in a directly earthed high-voltage (HV) network, raised potentials appear at places where they do not exist under normal operating conditions. The magnitudes of these potentials, as well as all of the potential differences relevant for the evaluation of safety conditions (touch and step voltages), are proportional to the current dissipating through the substation grounding system into the surrounding earth.

The grounding system of a HV substation, in general, justifies its existence only at the moment of an unbalanced fault, when high fault currents leave the power system and increased potentials appear at places exposed to human touch. To ensure protection against undesired consequences, such as the loss of human lives, burndown of grounding wires, and damage to sensitive telecommunication equipment entering the substation, it is necessary to design the corresponding grounding system of the planned HV substation. During the design procedure it is necessary to evaluate, as accurately as possible, the values of potentially dangerous voltages. When the calculated values of these voltages are within the prescribed limits, the designed grounding system is considered to satisfy the safety requirements and can represent an acceptable solution. Certainly, between several such solutions the most acceptable solution is the most cost-effective one.

The magnitudes of all potentials as well as of all potential differences, including touch and step voltages, are proportional to the part of the ground fault current dissipated from the substation grounding system into the surrounding ground. The favorable circumstance is that this current represents only a part of the total ground fault current. The rest of the ground fault current uses all other available return paths such as transmission line ground wires, cable sheaths, and other surrounding metal installations

Practical Methods for Analysis and Design of HV Installation Grounding Systems
ISBN 978-0-12-814460-2
https://doi.org/10.1016/B978-0-12-814460-2.00004-0

and their grounding connections. Thus they do not participate in the creation of any potential in the considered grounding system. Therefore when designing a substation grounding system it is of prime importance to evaluate, as accurately as possible, the part of the ground fault current that will be returned to the power system through the surrounding earth. Especially important is the fraction of the fault current dissipated into the soil through the substation grounding grid because only this element of the whole grounding system is designed and constructed exclusively for substation grounding purposes. Only after correctly determining this part of the total ground fault current it is possible to put the whole problem into a realistic framework and to arrive at a safe solution without excessive expenditure.

Evaluation of the distribution of the ground fault current for faults in substations supplied by transmission lines that can be treated as homogeneous through their entire length is performed by a relatively simple calculation procedure. These cases have been studied, systematized, and presented in Ref. [1] for various types of substations and for different types of transmission lines. Also in the case of cable feeding lines that can be treated as homogeneous along their entire length, adequate methods are developed in [2–5] and presented in Chapter 3.

However, these papers do not encompass all possible cases of importance for current engineering practice, i.e. do not encompass the cases when a feeding line has two or more sections with one or more mutually different relevant parameters. With the aim to considere this kind of problem several papers have been published [6–13].

The possibility of simplifying the problem and developing an analytical procedure has been seen in the reduction of the number of the elements of the electrical circuit spontaneously formed by this type of feeding line during a ground fault. To this purpose, formulae for an exact description of lumped parameter ladder circuits of any size (from one π to an infinite number of π-cells) by only one equivalent π, presented in Section 2.2.4, are used for presentation of transmission line ground wires and their grounding connections.

The method development is based on the usual approximations unavoidable at the design stage and by using the mentioned solution for lumped parameter ladder circuits. The finally derived analytical expressions enable evaluation of the ground fault current distribution when the feeding line represents a longitudinal combination of different sections, most frequently performed on a cable and an overhead section.

With permanent elevation of the ground fault current level and with an increasing deficiency of free and sufficiently large urban areas for the building of HV substations, correct determination of ground fault current distribution in various specific cases certainly gains in importance.

4.1.2 Problem Description

Large source substations supplying HV distribution networks to big cities require large areas and as a result are built outside of urban zones. Outgoing lines from the substations supply HV distribution substations (e.g., 110/10 kV), situated deep within urban areas, i.e., in the center of electric energy consumption. Each of the outgoing lines usually supplies two or three such substations in succession, forming the so-called radial distribution system. If these lines are made of cables along the whole of their lengths, from the source station to the supplied HV substation, calculation of ground fault current distribution for a fault in the supplied substation is considered in Chapter 3. However, since these lines pass through areas of different degrees of urbanization, we sometimes have a situation where a certain substation is supplied by a cable line as a continuation of an overhead line. Such a situation is symbolically depicted in Fig. 4.1.

When we have the situation depicted in Fig. 4.1 the problem of ground fault current distribution becomes somewhat more complicated. In such situations the current that appears due to the inductive coupling between faulted phase conductors on one side and: ground wire(s) and cable metal sheath(s) on the other side cannot be determined by formerly presented calculation procedures. One part of the current induced in cable sheath(s), at the place of discontinuity, continues its flow toward the supply sources in the power system through the ground. As a result, the reduction in fault current through the grounding system of the supplied substation is

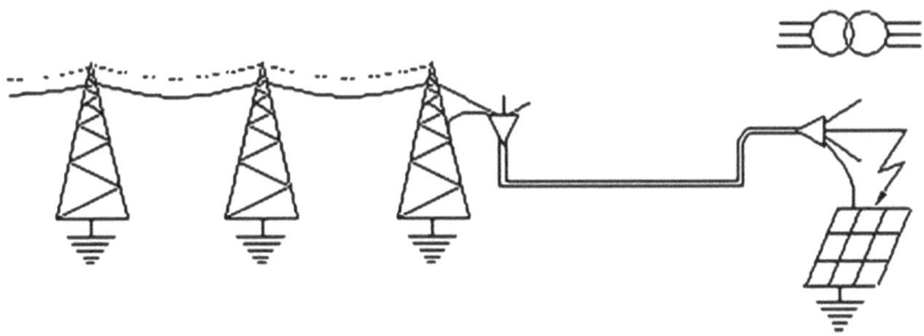

Figure 4.1 Feeding line consisting of a cable and an overhead section.

somewhat smaller than in the case where the feeding line is a cable along the whole of its length. On the other hand, this reduction is larger than in the case where the feeding line is an overhead line along its whole length. In practice, however, when determining the fault current distribution, such a line is sometimes treated either as a cable only or as an overhead line only. This may result in serious errors when estimating the achieved safety conditions within and in the vicinity of the supplied substation.

Because of the strong inductive coupling between phase conductors and cable sheath(s), most of the fault current can be injected into the ground at the place of discontinuity rather than at the fault location, i.e., in the supplied substation. When the place of discontinuity is in a transition station, this effect is described as the "fault application transfer effect," because a large part of the ground fault current is transferred to the transit HV substation [6].

Examples of such feeding lines are encountered in distribution networks with various voltage levels and as already mentioned are a consequence of various degrees of urbanization of the areas where the feeding line passes. Also such cases in current electric power engineering practice appear when a planned substation should be connected to a nearby existing line or when new substation *B* is connected by interpolation at point *D* to the existing line that previously connected stations *A* and *C* directly, as is symbolically depicted by Fig. 4.2.

Due to the situation depicted in Fig. 4.2, the feeding line in most cases is divided into one planned and one existing section. As a rule, the existing section is equipped with one steel ground wire, while the new section has one or two aluminum conductor steel-reinforced cables (ACSRs).

When a ground fault appears in the supplied substation, the lines presented by Figs. 4.1 and 4.2 form complex electrical circuits with a large number of conductively and inductively coupled elements.

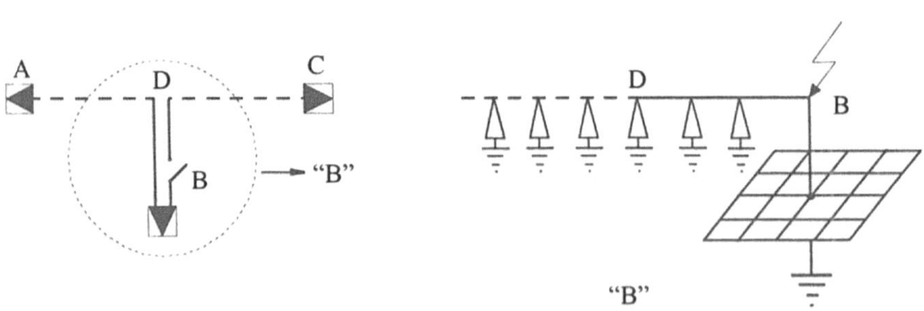

Figure 4.2 Feeding line consisting of two different sections [8].

4.2 SUBSTATION SUPPLIED BY A FEEDING LINE CABLE SECTION

4.2.1 Cable Having Uncovered Metal Sheath

In a grounded HV network the ground fault current distribution in phase conductors and the ground fault current distribution in its return paths represent two practically independent distributions that can be considered separately. An analysis of these distributions together would be more complicated and is unnecessary. Thus we will assume here that the ground fault current passing through the fault place is known from the prior ground fault studies and as such should not be a subject of our consideration.

Now, let us assume that the ground fault occurs at a distribution substation *B* supplied from substation *A* by a cable section of the feeding line consisting of an overhead section as well. We will start our consideration with the most general case, where the applied cable has an uncovered metal sheath and the overhead section has a ground wire. By using the following theoretical bases:

- method of symmetrical components,
- driving point technique,
- decoupling technique, and
- reduction of lumped parameter ladder circuits (Section 2.2.6).

In reality a very complex electrical circuit with numerous mutually coupled elements can be presented by the relatively simple equivalent circuit shown in Fig. 4.3.

The used notation has the following meaning:

F—fault place (substation *B*),

U_{ph}—auxiliary driving voltage source at fault place with the voltage equal to the phase voltage at the source station,

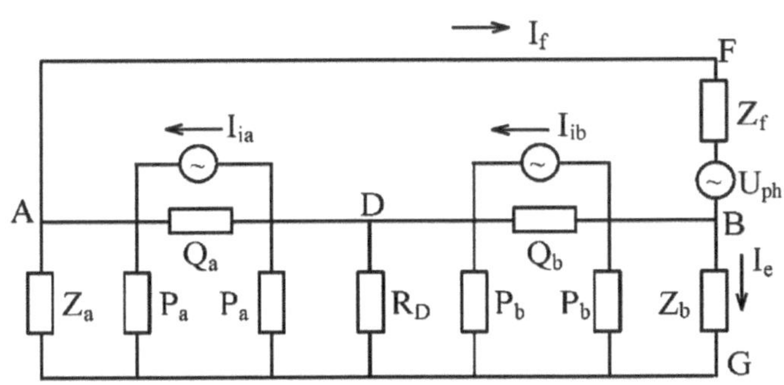

Figure 4.3 The equivalent circuit when a ground fault appears in the supplied substation [8].

I_f—total ground fault current,
I_e—current injected into the earth through the substation ground grid and through the connected external grounding electrodes (metal cable sheath[s], neutrals of the low-voltage (LV) network, metal pipelines, etc.) excluding the grounding effects of the considered line,
I_{ia}—current source that replaces the influence of the inductive coupling between the faulted phase conductor and the ground wire(s),
I_{ib}—current source that replaces the influence of the inductive coupling between the faulted phase conductor and the metal cable sheath(s),
Z_f—impedance of the electric power system at the fault place,
Q_a and P_a—impedances that make up the equivalent π-cell for the ladder network formed by the ground wire,
Q_b and P_b—impedances that make up the equivalent π-cell representing the cable sheath as grounding conductor,
$Z_a(Z_b)$—impedance of the grounding system of substation A (B) that does not incorporate the grounding effects of the considered feeding line,
R_D—footing resistance of the tower at the discontinuity place,
D—place of line discontinuity, and
G—remote ground.

When substation B has a grounded neutral point(s), the equivalent circuit from Fig. 4.3 should have an additional branch containing the zero sequence impedance of the local transformer and connects points F and B. According to the previously mentioned theory the elements representing the cable section as a grounding conductor, impedances Q_b and P_b, are determined by expressions (2.63) and (2.64), while the current induced in the metal sheath(s) of the cable section expressed through its reduction factor is determined by:

$$I_{ib} = (1 - r)I_f, \tag{4.1}$$

where r is the reduction factor of the cable section only (for three-core cables usually given as factory data).

The analytical expressions for the reduction factor for different types of cable lines and different cable formations are presented in Section 3.1.

For the part of the equivalent circuit representing the line overhead section, the impedances Q_a and P_a are determined by expressions (2.65) and (2.66), while the current induced in the overhead section of the line, expressed through its reduction factor, is determined by:

$$I_{ia} = (1 - r_0)I_f. \tag{4.2}$$

The reduction factor of the overhead line r_0 is, according to, e.g., Ref. [4], determined by:

$$r_0 = 1 - \frac{Z'_m}{Z'_{gw}}, \tag{4.3}$$

where Z'_{gw} is the self-impedance of the ground wire given per unit length and Z'_m is the mutual impedance between the ground wire and the faulted phase conductor per unit length.

Impedance Z'_{gw} is determined by Eq. (2.61), while impedance Z'_m is, according to, e.g., Ref. [4], determined by:

$$Z'_m = \omega_n \frac{\mu_0}{8} + j\omega_n \frac{\mu_0}{2\pi}\left(\frac{\mu_r}{4} + \ln\frac{\delta_n}{D_{gw}}\right), \Omega/\text{km}, \tag{4.4}$$

where D_{gw} is the distance between the faulted phase conductor and the ground wire (m), ω_n is the nominal angular frequency, $\omega_n = 2\pi f_n$, and δ_n is the earth penetration depth.

Since to estimate the safety conditions it is necessary to determine the largest possible value of the current I_e, impedance Z'_m should be determined under the assumption that the fault occurred on the phase conductor furthest from the ground wire.

The equivalent circuit does not comprise the "proximity effect between the cable and the ground grid" of supplied substation B. However, it is necessary to bear in mind that in many of these cases the mentioned effect has a small influence on ground fault current distribution. The main reason for this is the strong inductive coupling between the faulted phase conductor and the cable sheath(s). Due to this most of the ground fault current dissipates into the earth at the place of discontinuity, D, which in the case of any larger cable section is sufficient distance from supplied substation B.

Also in practical conditions the influence of impedance Z_a on ground fault current distribution is small and this impedance can be neglected ($Z_a \approx 0$). Thus to determine the ground fault current distribution we obtain the simplified equivalent circuit shown in Fig. 4.4.

In the circuit in Fig. 4.4, ΔI_i denotes the difference between the current I_{ib} and the current I_{ia}.

According to the equivalent circuit shown in Fig. 4.4 we can write a closed system of equations whose solution gives:

$$r_t = \frac{I_e}{I_f} = \frac{r_0(1-r)Z_D + r(Q_b + Z_D)}{\left(1 + \frac{Z_b}{P_b}\right)(Q_b + Z_D) + Z_b}. \tag{4.5}$$

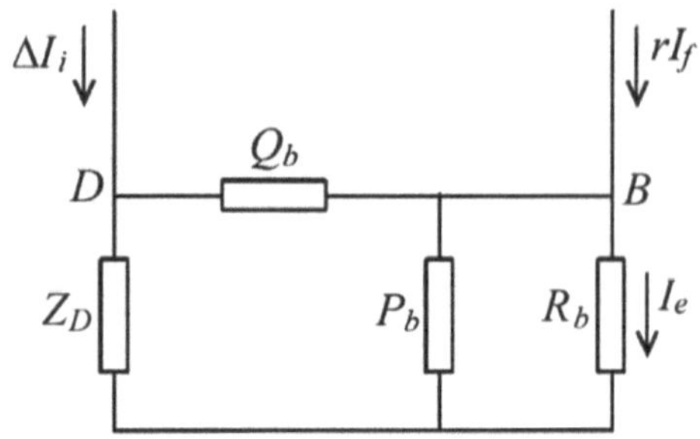

Figure 4.4 Simplified equivalent circuit [11].

In this expression, grounding impedance Z_D is determined by:

$$Z_D = \left(\frac{1}{R_D} + \frac{1}{Z_N} + \frac{1}{P_b}\right)^{-1}, \tag{4.6}$$

where impedance Z_N is, according to Eq. (2.69), determined by:

$$Z_N(Z_a = 0) = \frac{k^N - k^{-N}}{k^N + k^{-N}} Z_\infty, \tag{4.7}$$

where N denotes the span number of the overhead section.

The coefficient r_t represents the total feeding line reduction factor because it involves, besides the inductive influences, the grounding effect of the feeding line on the reduction of the part of the ground fault current injected into the earth, I_e.

Since, according to the definition of the line reduction factor, it should not involve the grounding effects of the feeding line, these effects should be excluded. We will eliminate them by adopting $Z_b = 0$. Then, according to Eq. (4.5), we obtain:

$$r_i = r + \frac{r_0(1 - r)Z_D}{Q_b + Z_D}. \tag{4.8}$$

Coefficient r_i includes only the influence of the inductive coupling between phase conductor and metal sheath(s) and phase conductor and ground wire(s) on the distribution of the ground fault current.

The second addend in Eq. (4.8) obviously represents the increment of the reduction factor in comparison to its value under the assumption that the feeding line is performed by cable(s) along the whole of its length. If we assume that the length of the cable section infinity increases, then the value of impedance Q_b tends to infinity ($Q_b \to \infty$) and the value of the reduction factor r_i tends to r ($r_i \to r$). Since under real conditions the reduction factor of a cable line is significantly lower than the reduction factor of an overhead line ($r \ll r_0$), it can be said that the value of r_i is situated between the two limiting values r and r_0 ($r < r_i < r_0$).

According to Eq. (4.8), factor r_i depends on the value of the grounding impedance at the place of discontinuity, Z_D. With a decrease of this impedance, if other relevant parameters remain unchanged, the value of the reduction factor r_i also decreases. If the value of this impedance becomes nearly negligible, the value of the reduction factor of the feeding line becomes practically equal to the reduction factor of the cable section, or $r_i \approx r$. On the other hand, with an increase of impedance Z_D, if other relevant parameters remain unchanged, the value of r_i increases and tends to the value of the reduction factor of the overhead line, r_0.

4.2.2 Special Cases

The presented equivalent circuits and the given analytical expressions are valid for the general case but can be easily modified for various special cases that can be encountered in practice.

When the length of the overhead section is such that we can consider it infinite from the point of view of the grounding effects, instead of Eq. (4.6), we can use:

$$Z_D = \left(\frac{1}{R_D} + \frac{1}{Z_\infty} + \frac{1}{P_b} \right)^{-1}, \tag{4.9}$$

where Z_∞ denotes the grounding impedance of the infinite overhead section.

A line overhead section with N spans can be considered infinite from the point of view of the grounding effects if its length is longer than, or at least equal to, its active length. The active length of an overhead line (ground wire) can be estimated on the basis of Eq. (2.80).

When the place of discontinuity is a transition substation, impedance Z_D is equal to the grounding impedance of this substation, or $Z_D = Z_{gt}$, where Z_{gt} denotes the grounding impedance of the transit substation.

In the case of a steel ground wire the reduction factor r_0 has a large value ($r_0 \approx 0.95$) so that expressions (4.5) and (4.8) can be simplified by introducing the approximation $r_0 \approx 1$. Also when the overhead section is without a ground wire, then we have $r_0 = 1$ and $Z_D \approx P_b$, so that, instead of Eq. (4.8), we have:

$$r_i = \frac{rQ_b + P_b}{Q_b + P_b}. \tag{4.10}$$

Finally, when the length of the cable section tends to its active length, according to Eqs. (2.63) and (2.64), Q_b tends to infinity and P_b to $Z_{c\infty}$, so that, instead of Eq. (4.8), we have:

$$\lim_{\substack{Q_b \to \infty \\ P_b \to Z_{c\infty}}} r_i = \lim_{\substack{Q_b \to \infty \\ P_b \to Z_{c\infty}}} \frac{rQ_b + P_b}{Q_b + P_b} = r. \tag{4.11}$$

On the basis of Eq. (4.11) we can write:

$$r_i(L \geq L_a) = r. \tag{4.12}$$

As can be seen from Eq. (4.12), the reduction factor of a cable section longer than its active length becomes equal to the reduction factor of the feeding line that is performed along its whole length as a cable line.

By using the corresponding values for the relevant parameters the active cable length in each concrete case can be estimated by using the transcendental Eq. (2.94).

4.2.3 Cable Having Covered Metal Sheaths

This type of cable does not have grounding properties, i.e., does not dissipate the ground fault current through its metal sheath into the surrounding earth, because its outer sheath is made of insulated materials. By taking this into account it is necessary to modify the circuit shown in Fig. 4.4. Impedance P_b is omitted ($P_b = \infty$), and for impedances Q_b and Z_D the following relations are valid:

$$Q_b = LZ'_{sh}, \tag{4.13}$$

$$Z_D = \left(\frac{1}{R_D} + \frac{1}{Z_N}\right)^{-1}, \tag{4.14}$$

where L is the length of the cable section (km) and Z'_{sh} is the self-impedance of the cable sheath(s) per unit length (Ω/km).

Then, on the basis of Eqs. (4.8) and (4.13), we have:

$$r_i = r + \frac{r_0(1 - r)Z_D}{LZ_{sh} + Z_D}. \tag{4.15}$$

In common cases, when the overhead section regarding the grounding effects may be considered as infinitely long, instead of Z_N we have Z_∞ in Eq. (4.14), and when a place of a discontinuity is a transition station, then $Z_D = Z_{gt}$.

The last case can be encountered in HV (e.g., 110 kV) distribution networks. The real values of L and Z_{gt} in these networks are such that the condition $Z_{gt}/LZ_{sh} < 0.05$ is satisfied in almost all cases and instead of Eq. (4.15) the following approximation can be used: $r_i \approx r$.

If the discontinuity place is on the line itself, the reduction effect of this type of cable is less pronounced, because the grounding impedance Z_D in that case is significantly larger ($P_b = \infty$). Thus in some cases it could be more economical to undertake measures leading to a lower value of R_D (Eq. 4.14) than measures leading to a lower value of the grounding resistance of the supplied substation ground grid. Certainly, a separate consideration is needed in each concrete case.

In medium-voltage (MV) distribution networks it is common that several substations are successively supplied through the same cable section. This situation is symbolically depicted by Fig. 4.5.

However, since the lumped parameter ladder circuit formed by the cable sheath(s) and the grounding electrodes of the supplied stations is analogous to the ladder circuit formed by the overhead ground wire, this network can also be represented by only one equivalent π-cell. In that case, Q_b and P_b are also defined by Eqs. (2.65) and (2.66). Only instead of Z_S and R we now have the average value of the self-impedance of the cable sheath(s) between the two nearby substations and the average value of the ground resistances of the substations immediately supplied by the cable section. The ground fault current distribution when a ground fault occurs in substation B is symbolically depicted by Fig. 4.6.

The used notation has the following meaning:

A, B, and C—substations supplied by the cable section of the feeding line,

I_f—total ground fault current,

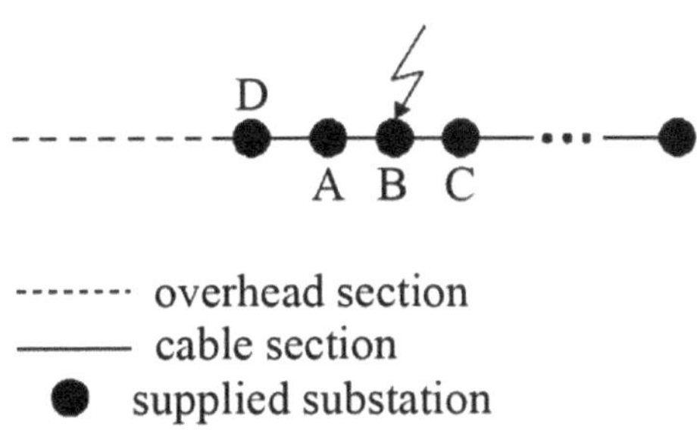

Figure 4.5 Cable section supplying several medium-voltage/low-voltage (MV/LV) substations [11].

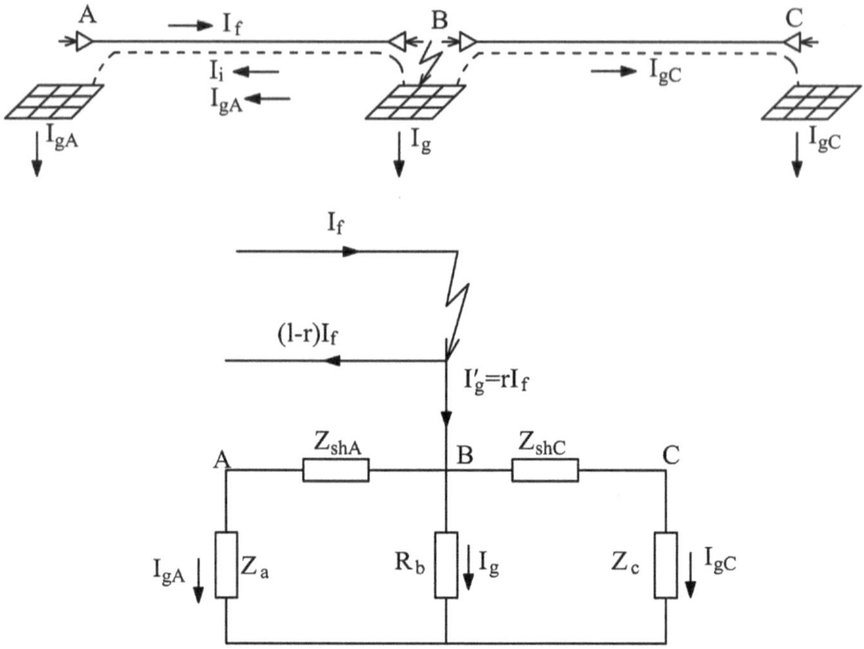

Figure 4.6 Ground fault current distribution.

I_i—current appearing in the cable sheath solely as a consequence of the inductive coupling between the cable sheath(s) and the faulted phase conductor,

I'_g—current injected into the earth through the grounding system of substation B,

I_g—current injected into the surrounding earth through the ground electrode of substation B,

I_{gA} (I_{gC})—current appearing in the cable sheath(s) solely as a consequence of the conductive connection with the grounding system of substation A (C),

Z_a, (Z_b)—grounding system impedances of substation A (C) without the grounding effects of cable section AB (CB),

Z_{shA} (Z_{shC})—self-impedance of the cable sheath(s) from substation A (C) to B,

R_b—resistance of the grounding electrode of substation B, and

r—reduction factor of the feeding line determined by Eq. (4.8).

The previous analysis showed that the reduction of current I_g in the case of a feeding line consisting of a cable and an overhead section depends on many factors such as the cable section length, the type of the applied cable (with an insulated or with an uncoated outer sheath), grounding impedance of the point of discontinuity, the reduction factor of the overhead section, and the reduction factor of the cable section seen separately.

On the basis of Eq. (4.15), it is important to note that the reduction factor for a sufficiently long cable section becomes equal to the reduction factor of the feeding line that is cable on the whole of its length. The same is true when, independently of the cable section length, the grounding impedance at the place of discontinuity Z_D is negligibly small.

4.2.4 Quantitative Analysis

4.2.4.1 Feeding Line Consisting of Cable and Overhead Section

The analysis is performed with the aim of assisting designers in developing an engineer's feeling for the influences of certain relevant parameters. The necessary numerical data are adopted under the assumptions and approximations unavoidable at the design stage.

At first we shall assume that the MV substation is directly fed by a three-core cable with an uncoated outer sheath, which is connected at the other end to an overhead line without any ground wire (Z_N and R_N are infinitely large). This situation is quite common in mid-voltage distribution networks with oil-filled, three-core cables. After the discretization of the cable sheath based on the following data: outer diameter $d = 44$ mm and experimentally determined $Z_m/Z_s \approx 0.8$ and $Z_s = (0.0007 + j0.002)\ \Omega$/km, the calculations performed on the basis of the presented analysis give the results that are graphically presented in Fig. 4.7.

As can be seen, when the cable length L_c increases, it is evident that the reduction factor r_i tends to the value it would have if the cable were laid along the whole length of the feeding line.

Now we shall assume that one cable section immediately supplies a 110/10 kV distribution substation, while, at the other end, the cable section is connected to the overhead section of the feeding line in the nearest transit substation. The cable has lead sheath and its external and internal diameters are equal, 54.6 and 51.4 mm, while the value of the reduction factor of the cable, according to the manufacturer's data, is equal to 0.22. For the overhead part of the line we assume that in one case it is equipped with a steel ground wire, so that it can be adopted, $r_0 = 1$, while in the other case we have a ground wire ACSR 95/55 mm^2 with a reduction factor $r_0 = 0.675$. The calculation results of the feeding line

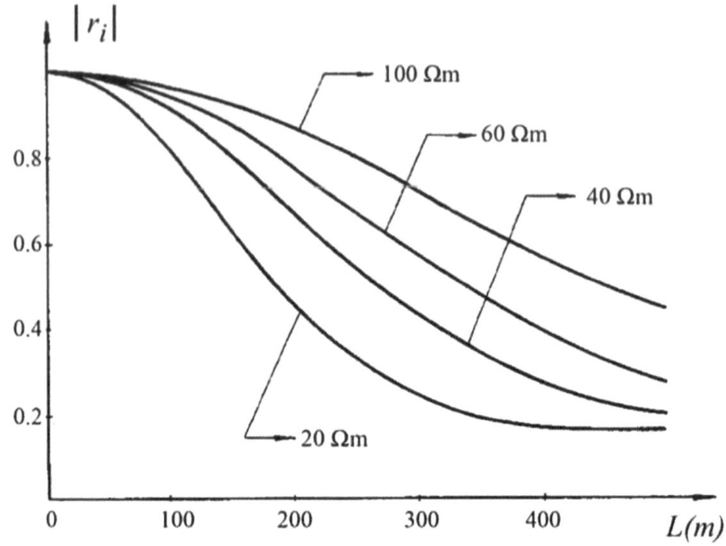

Figure 4.7 Reduction factor a cable line with uncoated sheath in the case: $r_0 = 1$, $Z_N = \infty$, $R_D = \infty$ [8].

reduction factor versus cable section length for the first and second cases are shown in Figs. 4.8 and 4.9, respectively.

If we assume that the length of the cable section is 3 km (the average length of the cable lines in the 110 kV network of Belgrade, Serbia, is

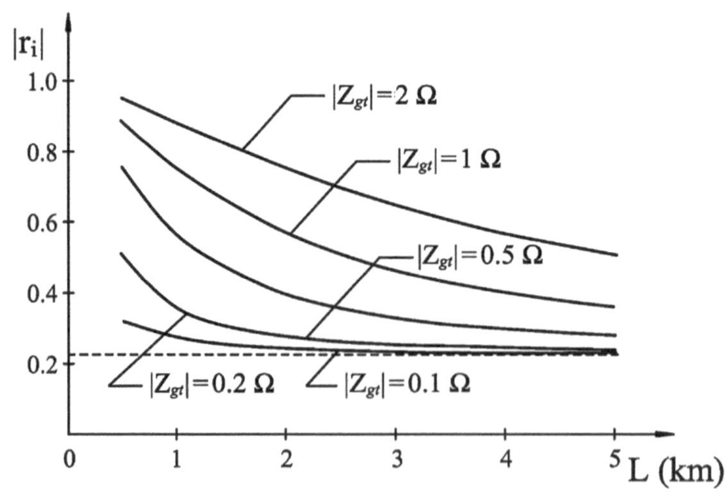

Figure 4.8 Reduction factor in the case of a cable with an insulated outer sheath and $r_0 = 1$ [11].

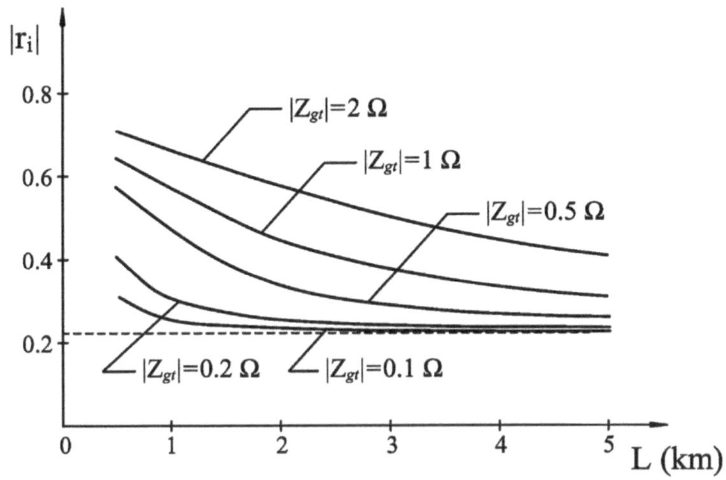

Figure 4.9 Reduction factor in the case of a cable with an insulated outer sheath and $r_0 = 0.675$ [11].

approximately 4 km), the increase in the reduction factor $|r_i|$ in comparison to $|r|$ is for the first case (Fig. 4.8) 6.4%, 13%, 49%, 110%, and 195%, while for the second case (Fig. 4.9) it is 4%, 10%, 32%, 71%, and 128% for $|Z_{gt}| = 0.1$, 0.2, 0.5, 1, and 2 Ω, respectively. On the basis of the diagrams given in these figures it is not difficult to see the following: when the value of impedance Z_{gt} is lower, the influence of the type of ground wire or the value of r_0 is lower. Then, when the cable section is longer, the influence of the type of ground wire of the overhead section or the value of r_0 is also lower.

The effective value of the grounding impedance of substations in 110 kV distribution networks higher than 0.1 Ω ($|Z_{gt}| > 0.1$ Ω) is rare in practice and may be expected only in cases of unfavorable (high) soil resistivity. However, according to the presented diagrams, one can gain an insight into the value of the reduction factor when impedance Z_{gt} takes unrealy high values. Also it is important to note that the greater part of the ground fault current is dissipated into the earth through the grounding system of the transit substation.

The considered cases are not too rare in contemporary electric power distribution practice, especially in suburban areas where there are different degrees of urbanization. The calculations are relatively simple and take into account all of the relevant data available at the design stage. In other words, the presented method is developed on the basis of the approximations identical to those that we are otherwise forced to adopt, because of the uncertainty of some relevant data, at the design stage.

On the basis of the given results one can obtain the following conclusions. Estimation of the safety conditions based on the cable reduction factor r would give somewhat more favorable results, i.e., lower values of the touch and step voltages in comparison with the actual ones. This means that it could happen that the prescribed safety criteria in some cases could be treated as satisfied, although this is not the case in reality. This should be stressed because the standard measurements of safety conditions, which are mandatory before a newly built substation is put in operation, usually do not involve verifications of the reduction factor value of the feeding line.

4.2.4.2 Feeding Line Consisting of Different Overhead Sections

Let us assume that we have the situation shown in Fig. 4.2 (substation A is the only supply substation), where the new section of 110 kV line is equipped with two ground wires, ACSR 95/55 mm^2, and the existing one has one ground wire, steel 50 mm^2. Also we shall assume that the length of all line spans is 250 m and that the existing line section can be considered as infinite from the standpoint of grounding effects. The resistances per unit length of steel and ACSR ground wire are 3.5 and 0.299 Ω/km, respectively, while geometric mean radii are 3.49 and 6.23 mm, respectively. The line configuration for different sections is shown in Fig. 4.10.

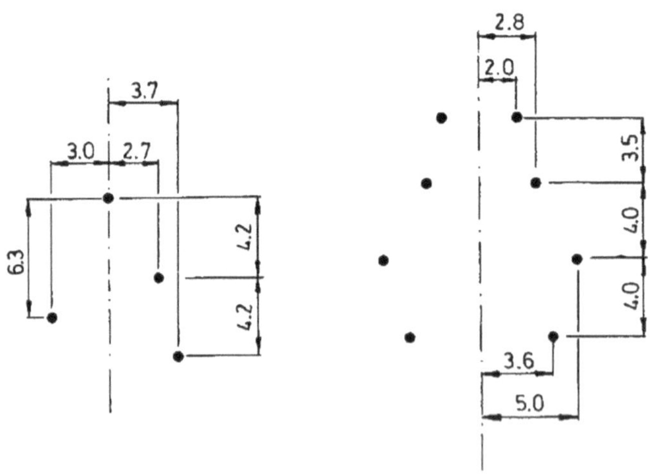

Figure 4.10 Disposition of line conductors of two different sections (dimensions in meters) [8].

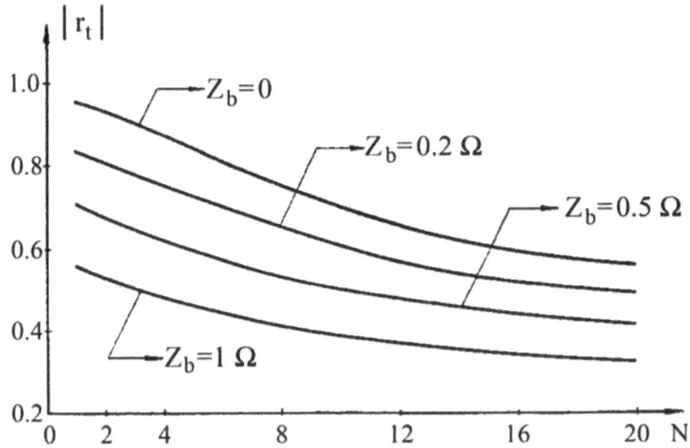

Figure 4.11 Influence of number of spans with ACSR ground wires for different values of Z_b [8].

On the basis of the presented analytical expressions the performed calculations give results graphically presented in Fig. 4.11.

From the given diagrams it is clear that the considered effect is expressed more when the need is greater, i.e., when the grounding impedance of the supplied substation, Z_b, is larger.

As shown, the presented method at the design stage enables a necessary evaluation of the ground fault current distribution in substations supplied by the feeding line consisting of sections with mutually different parameters. By applying this method, the number of practically possible cases that can be solved by a relatively simple calculation procedure, or without the use of computers, is extended.

4.3 MEASUREMENTS OF REDUCTION OF THE GROUND GRID POTENTIAL RISE

4.3.1 A Substation Supplied by an Overhead Line

4.3.1.1 Introduction

In some specific and for solving difficult cases the most acceptable solution can be obtained by undertaking the measure for reducing the ground grid potential rise. The usual reason for this is too high a potential transferred to the nearest supplied MV/LV substation, or to the incoming telecommunication circuits, railway lines, gas pipelines, etc.

In general, ground fault currents return to the generating sources in a power system through the earth and neutral conductors. Potential of a substation ground grid, as well as all potential differences relevant for safety conditions (step and touch voltages), appears as a consequence of only one fraction of the total ground fault current. This is the part of the ground fault current that is dissipated from the substation grounding system into the surrounding earth. Therefore when designing a substation grounding system it is necessary to evaluate as precisely as possible this part of the total ground fault current.

However, in cases where conditions for improving the substation grounding problem are not favorable (high short-circuit level, high soil resistivity, a confined yard foreseen for the substation, vicinity of public buildings, etc.) it may happen that this is not sufficient, e.g., Ref. [6]. In such cases the solution can require a special measure that leads to an overall reduction in the values of potentials and potential differences within the substation and in its vicinity by reducing the portion of the ground fault current that will be dissipated through the grounding system of the supplied substation.

Practical implementation of the measures for changing ground fault current distribution, in the case of overhead lines, or the so-called counterpoise, are usually only described in the corresponding technical standards, while the method for calculating the achieved effects is omitted.

An illustration of the practical application of the counterpoise is shown in Fig. 4.12, where source station A supplies HV distribution substation B through an overhead line. As can be seen, the counterpoise can be described as burying a bare copper wire of certain length immediately under the overhead feeding line and in the vicinity of supplied substation B. The copper wire is connected to this substation ground grid and to the ground electrodes of a few close transmission towers.

The used notation has the following meaning:

A (B)—source (supplied) substation,

$Z_a(Z_b)$—impedance of the grounding system of substation $A(B)$,

D—place of discontinuity,

d_{cc}—diameter of the counterpoise conductor (copper wire), and

h—depth of the buried counterpoise conductor.

As can be seen, the counterpoise measure is relatively easy for practical implementation. However, because of the complexity of the electrical circuit formed during a ground fault, the same cannot be said for the calculation of the achieved effects. The current dissipated into the surrounding earth produces a large potential field around the substation ground grid. Since the counterpoise conductor is passing through the area involved by this field, a mutual interference between the ground grid and

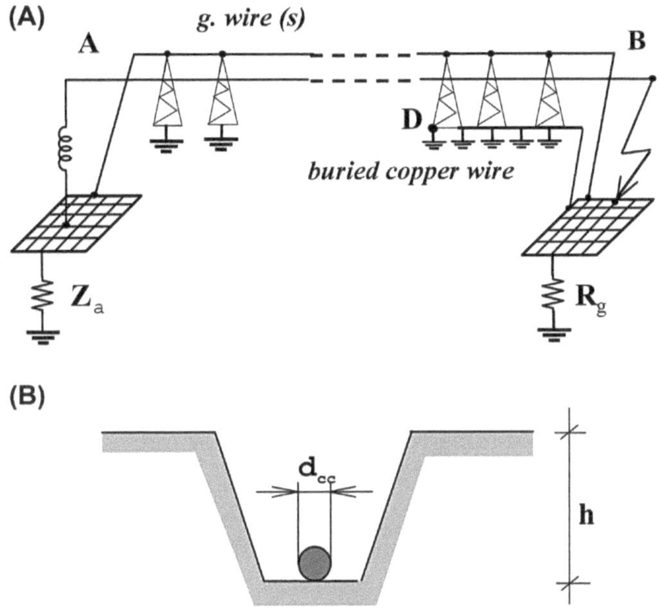

Figure 4.12 Practical application of the counterpoise. (A) Feeding line. (B) Copper wire buried along the line track [10].

the counterpoise conductor in dissipating the ground fault current into the earth, known as the "proximity effect," is unavoidable. Moreover, one portion of the ground fault current appears in the counterpoise conductor as a consequence of the inductive coupling with the feeding line conductors. Thus, under practical conditions, the magnitude of the current diverted from the substation grounding grid by a counterpoise conductor is a very complex function of the self- and mutual impedances of the overhead line conductors and counterpoise conductor, substation grounding impedance, overhead line tower footing resistances, proximity effect between the ground grid and the counterpoise conductor, and the length of the counterpoise conductor. Therefore certain idealizations and simplification of the real physical model were unavoidable to develop a simple and for practical purposes sufficiently accurate method [10].

This method is directly (without using a computer) applicable for estimation of substation ground grid current reduction achieved by using the counterpoise. It takes into consideration all essential factors and parameters with reasonable approximations or idealization of the actual physical model. For solving especialy difficult cases, reduction of the substation grid current achieved by using the counterpoise conductor can be the only applicable solution. Also in other cases it is not the only, but can be the most economical, solution. As a result, it can be considered a useful alternative solution in cases where other solutions also satisfy the safety criteria, but are too expensive.

4.3.1.2 Description of the Problem

Let us assume that supplied substation B is an outdoor HV distribution substation (Fig 4.12). As is well known, the grounding problem of such substations can be very pronounced and as such deserves special consideration. The difficulties originate from the immediate urban or suburban environment, and the feeding line constructed overhead, as is well known, has high reduction factor values.

The number of outgoing MV cable lines, with uncovered or insulated metal sheath(s), can be relatively small, especially in the critical moment of putting a newly built substation into operation. As a result, the possibilities of attaining required safety conditions, especially on consumer premises (in the other words, sufficiently low values of transferred potentials), can be very limited, especially in the case of high local soil resistivity or high level of ground fault currents.

Obviously, the counterpoise conductor represents an additional grounding conductor, which provides an additional return path for the ground fault current. Thus the problem is: how can the portion of the ground fault current dissipating into the earth through the substation grounding system be determined, as well as its part, current dissipated into the earth only through the substation ground grid?

The previously defined problem is so complex that certain idealizations of the actual physical model are unavoidable. Because of the many factors and parameters involved, such an approach is necessary at the very beginning, i.e., when we have to collect all necessary input data. An illustrative example of this is the value of the equivalent soil resistivity along the counterpoise conductor. It should be mentioned that the soil through which the counterpoise conductor passes is normally heterogeneous and an exact determination of its equivalent resistivity in urban conditions is not practically possible. Due to this, we have to assume that the soil is homogeneous and of a resistivity somewhat higher than that approximately assessed at its highest value.

Thus in the procedure of developing a necessary method, reasonable assumptions and approximations can be used leading to an accuracy sufficient for solving practical problems [14–16].

We will consider the case where the grounding problem of HV substations appears in the most expressed form because the need for the considered measure is then greatest. Certainly, this is the case where instead of a grounding system we have only the substation ground grid, e.g., the case where all outgoing MV lines are foreseen as overhead ones and without ground wires.

4.3.1.3 The Equivalent Circuit of the Ground Fault Current Distribution

According to the situation depicted in Fig. 4.12, it is not difficult to imagine that during a ground fault, a complex electrical circuit, composed of many inductively and conductively coupled elements, is formed. With the aim of simplifying the actual electrical circuit, we will make the following assumptions:

- The contribution of the towers' footing electrodes to dissipate the fault current along the line section of the counterpoise is negligible.
- The counterpoise conductor is conductively connected to the overhead line ground wire only at its ends, while other connections (the towers between these points) are disregarded.
- The currents in the neutral conductors (ground wire and counterpoise conductor) have negligible influence on the total ground fault current, I_f, and the current induced in the ground wire.
- The last assumption can be considered as real if the following criterion is satisfied: length of the counterpoise conductor L_{cc} is significantly smaller than the feeding line length L ($L_{cc} < 0.2L$).

On the basis of these assumptions and approximations and by using:

- the method of symmetrical components,
- the decoupling technique, and
- the analytical expressions for reduction of a uniform ladder circuit.

The grounding system shown in Fig. 4.12 can be presented by the relatively simple equivalent circuit shown in Fig. 4.13.

The used notation has the following meaning:

I_f—ground fault current component passing through the grounded neutrals in the supply network (known from the former ground fault current analysis),

ΔI_i—difference between the currents induced in the neutral conductors of the counterpoise section and the current induced in the ground wire of the rest of the line,

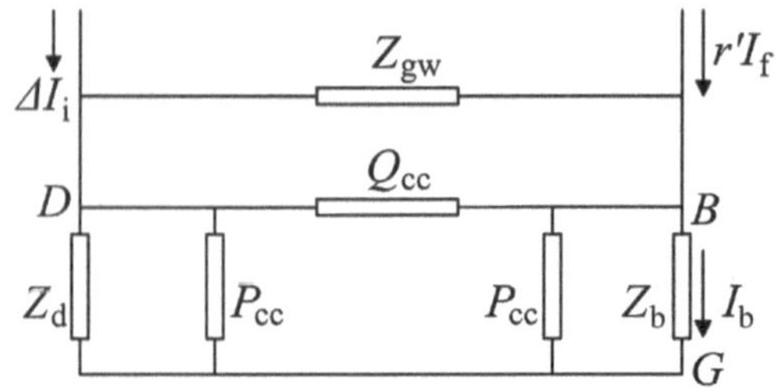

Figure 4.13 Ground fault current distribution in the grounding system of substation *B* [10].

I_g—fault current dissipated into the ground through the substation ground grid,
r'—reduction factor of the feeding line, which includes the inductive influence of the counterpoise conductor,
Z_{gw}—self-impedance of the ground wire of the section with a counterpoise,
Z_d—grounding impedance of the ground wire seen from *D* toward substation *A* (Fig. 4.12),
Q_{cc} and P_{cc}—impedances that make the equivalent π-cell for the lumped parameter ladder circuit formed by discretization of the counterpoise conductor,
R_g—resistance of the ground grid of the supplied substation,
D—end point of the counterpoise conductor, and
G—remote ground.

On the basis of the equivalent circuit of Fig. 4.13, one can determine the current dissipated through the substation ground grid, I_g. However, prior to this it is necessary to know all the elements of this circuit.

The total current appearing in the neutral conductors (ground wire and counterpoise conductor) as a consequence of their inductive coupling with the faulted phase conductor is larger than the current appearing due to the same reason in the ground wire of the rest of the line. The difference between these two currents at the point of discontinuity, *D*, has to continue its flow toward the power system solely through the ground. This current is denoted as ΔI_i in Fig. 4.13 and is determined by:

$$\Delta I_i = \left(\frac{Z'_m}{Z'_S} - \frac{Z'_{mC}}{Z'_{SC}}\right) I_f = (r_0 - r_c) I_f, \tag{4.16}$$

where Z'_S is the self-impedance of the transmission line ground wire(s) per unit length, Z'_m is the mutual impedance between the faulted phase conductor and the ground wire(s) per unit length, Z'_{SC} is the self-impedance of the neutral conductors per unit length, Z'_{mC} is the mutual impedance between the faulted phase conductor and the neutral conductors per unit length, r_c is the reduction factor of the line section of counterpoise application, and r_0 is the reduction factor of the rest of the line.

The reduction factor of the feeding line section of the counterpoise can be determined by elimination of the parts of the total ground fault current appearing in the neutral conductors as a consequence of their conductive coupling with the grounding electrodes at the counterpoise conductor ends. This can be done by assuming that R_g and Z_d are negligibly small ($R_g = 0$ and $Z_d = 0$). Then, the equivalent circuit presented in Fig. 4.13 can be substituted by the simpler one shown in Fig. 4.14.

The used notation has the following meaning:

Z_{cc}—self-impedance of the counterpoise conductor,

Z_m—mutual impedance between the ground wire and the faulty phase conductor,

Z_{m1}—mutual impedance between the counterpoise conductor and the faulted phase conductor, and

Z_{m2}—mutual impedance between the ground wire and the counterpoise conductor.

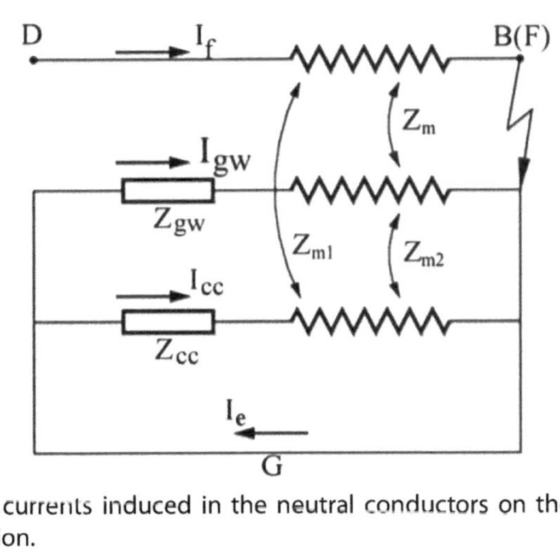

Figure 4.14 The currents induced in the neutral conductors on the section of counterpoise application.

On the basis of the equivalent circuit in Fig. 4.14, one can write the following system of equations:

$$\begin{aligned} & Z_{gw} I_{gw} + Z_{m2} I_{cc} + Z_m I_f = 0, \\ & Z_{cc} I_{cc} + Z_{m2} I_{gw} + Z_{m1} I_f = 0, \\ & r_c = 1 + \frac{I_{gw} + I_{cc}}{I_f}. \end{aligned} \tag{4.17}$$

By solving this system of equations one obtains that the reduction factor of the line section with the applied counterpoise, seen separately, is:

$$r_c = 1 - \frac{Z_{m1}(Z_{gw} - Z_{m2}) + Z_m(Z_{cc} - Z_{m2})}{Z_{gw} Z_{cc} - 2Z_{m2}^2}. \tag{4.18}$$

Since we need the largest/critical value of current I_g, it should be noted at this point that the choice of the faulted phase conductor has to be done carefully. The phase conductor at the greatest mean geometrical distance from the neutral conductors has not to be critical in each considered case. Thus calculation of the reduction factor should be performed for each phase conductor separately with the aim of obtaining the largest one. In most practical cases, in the rest of the line the phase conductors are transposed and should be treated as such in the determination of the reduction factor, r_0.

Then, the grounding grid resistance can, in the majority of practical cases, according to Ref. [15], be determined by the following formula:

$$R_g = \rho \left[\frac{1}{l_g} + \frac{1}{\sqrt{20A}} \left(1 + \frac{1}{1 + h_g \sqrt{20/A}} \right) \right], \tag{4.19}$$

where R_g is the resistance of the substation grounding grid (Ω), ρ is the equivalent soil resistivity (Ωm), l_g is the total length of the buried conductor (m), A is the area covered by the grounding grid (m^2), and h_g is the depth of the grounding grid (m).

This expression can be used with reasonable accuracy for grid depths between 0.25 and 2.6 m.

The uniform ladder circuit obtained by the discretization of the counterpoise conductor is presented in Fig. 4.13 by using equivalent impedances Q_{cc} and P_{cc}. On the basis of the analogy between this conductor and the uncovered cable sheath as a grounding conductor, these impedances are also determined by Eqs. (2.63) and (2.64), i.e., by:

$$Q_{cc} = \frac{k^{L_{cc}} - k^{-L_{cc}}}{2} Z_{cc\infty}, \tag{4.20}$$

$$P_{cc} = \frac{k^{L_{cc}} + k^{-L_{cc}}}{k^{L_{cc}} - k^{-L_{cc}}} Z_{cc\infty}, \tag{4.21}$$

where:

$$k = 1 + \frac{Z_{cc\infty}}{R_{cc}}, \tag{4.22}$$

and:

$$Z_{cc\infty} = \frac{Z_{cc}}{2} + \sqrt{R_{cc} Z_{cc} + \frac{Z_{cc}^{2}}{4}} \approx \sqrt{R_{cc} Z_{cc}}, \tag{4.23}$$

whereas L_{cc} represents the total counterpoise conductor length expressed in relation to the length of 1 m.

Variables Z_{cc} and R_{cc} in Eq. (4.23) denote discrete parameters of the segment of the counterpoise conductor of length 1 m. Resistance R_{cc} is, according to Ref. [15], determined by:

$$R_{cc} = \frac{\rho}{\pi m} \ln \frac{L_{cc}}{\sqrt{d_{cc} h}}, \ \Omega, \tag{4.24}$$

where d_{cc} is the diameter of the counterpoise conductor (m), h is the counterpoise conductor burial depth (m), and m is the length of 1 m.

Impedance Z_{cc} is determined by:

$$Z_{cc} = Z'_{cc}(\Omega/\mathrm{m}) \times 1 \text{ m}, \tag{4.25}$$

where Z'_{cc} is the the self-impedance of the buried counterpoise conductor per meter and can be determined by using Eq. (2.55).

Impedance Z_{gw} (Fig. 4.13 and 4.14) is determined by:

$$Z_{gw} = Z'_{gw}(\Omega/m) \times L_{cc}, \tag{4.26}$$

where Z'_{gw} is the self-impedance of the ground wire per unit length determined by Eq. (2.61).

Finally, grounding impedance Z_d is, according to expression (2.69), determined by:

$$Z_d \approx Z_d(Z_A \approx 0) = \frac{k^N - k^{-N}}{k^N + k^{-N+1}} Z_{\infty}. \tag{4.27}$$

The coefficient of the current distribution k is determined by:

$$k = 1 + \frac{Z_{\infty}}{R_t}, \tag{4.28}$$

$$Z_{\infty} = \frac{Z_S}{2} + \sqrt{R_t Z_S + \frac{Z_S^2}{4}}, \tag{4.29}$$

where N is the number of spans of the feeding line from the point of discontinuity, D, to source substation A, Z_S is the self-impedance of the ground wire(s) for average span (Ω), and R_t is the expected average value of the towers' footing resistance (Ω).

Since the analytical expressions for determination of each of the elements of the circuit of Fig. 4.13 are known, the current I_g can be determined by the principle of superposition, i.e., by:

$$I_g = I_g(r_c I_f) + I_g(\Delta I_i), \tag{4.30}$$

where $I_g(r_c I_f)$ is the component of current I_g caused only by the ideal current source, $r_c I_f$ (Fig. 4.13), and $I_g(I_i)$ is the component of current I_g caused only by the ideal current source, ΔI_i (Fig. 4.13).

In the majority of practical cases, the lengths of transmission lines are significantly larger than their, from the standpoint of grounding effects, active length. Thus in many practical cases, grounding impedance Z_d is independent of the line length (N) and instead of Eq. (4.27) one can use the simpler expression Eq. (4.29).

The degree of reduction of current I_g increases if the length of the counterpoise conductor (buried copper wire) increases. However, this reduction cannot be limitless. Any increase in the counterpoise conductor length in excess of certain limit length will not cause further reduction in the value of current I_g. This, the so-called active length of long grounding conductors, can be, on the basis of Eq. (2.94), estimated by:

$$L_a(\varepsilon = 0.05) \approx \frac{3m}{\ln\left(1 + \sqrt{\dfrac{\pi |Z_{cc}| m}{\rho \ln \frac{L_a}{\sqrt{h d_{cc}}}}}\right)}. \tag{4.31}$$

With the aim of obtaining the maximal possible effect of the counterpoise application, the length of the counterpoise conductor should be approximately equal to its active length or somewhat larger. When this condition is satisfied, impedance P_{cc} becomes equal to $Z_{cc\infty}$ ($P_{cc} = Z_{cc\infty}$), while impedance Q_{cc} becomes infinitely large ($Q_{cc} \rightarrow \infty$). Also in this case the current produced by ideal current source ΔI_i through the grounding resistance R_g becomes negligible. Thus the equivalent circuit of Fig. 4.13 obtains the simpler form shown in Fig. 4.15.

The used notation has the following meaning:

I_{gw}—current generated by current source $r_c I_f$ in the ground wire(s),

I_{cc}—current generated by current source $r_c I_f$ in the counterpoise conductor,

Z'_{gwb}—grounding impedance of the ground wire(s) seen from substation B, and

Z_{cp}—grounding impedance of the counterpoise conductor, including proximity effect with the ground grid.

Impedances Z'_{gw} and Z_{cp} in Fig. 4.15, according to the previous simplifications of the actual electrical circuit, are:

$$Z'_{gw} = Z_{gw} + Z'_{d}, \tag{4.32}$$

where:

$$Z'_{d} = \frac{Z_{\infty}\, Z_{cc\,\infty}}{Z_{\infty} + Z_{cc\,\infty}} \frac{I_{gw} + \Delta I_i}{I_{gw}}. \tag{4.33}$$

In real conditions, the length of the counterpoise conductor and dimensions of the grounding grid have such values that influence the potential field of the grounding grid at the end of the counterpoise conductor, D, and can be neglected. Also on the basis of the circuit shown in Fig. 4.13 it is clear that current $r_c I_f$ is most intensively injected into the ground at the beginning of the counterpoise conductor, while current ΔI_i is most intensively injected into the earth at its end. According to this, it can be concluded that the proximity effect influences only the distribution of current $r' I_f$ (Fig. 4.13).

The presented procedure is focused on the grounding effects of the counterpoise conductor and gives the method of their determination, but under the assumption that the proximity effect between the ground grid

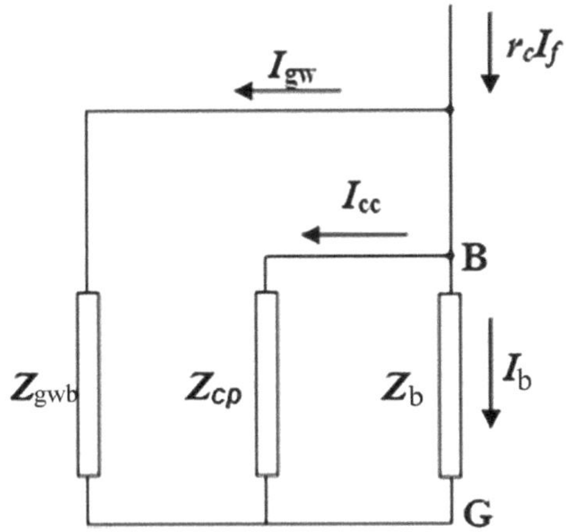

Figure 4.15 The simplified equivalent circuit [10].

and the counterpoise conductor is negligible. However, if more accurasy is desired this effect can be approximately taken into account through the value of impedance P_{cp} by using Eq. (6.21).

4.3.2 A Substation Supplied by a Cable Line

4.3.2.1 Introduction

Development of the method that took into consideration the favorable influence of the metal installations surrounding feeding line made the grounding problem of HV distribution substations supplied by cable lines very easy from the standpoint of necessary expenditure (Section 6.9). However, under the assumption of extremely unfavorable conditions (high value of soil resistivity, high value of the ground fault current, small number of metal installations surrounding the feeding line, small number of external electrodes, confined yard for substation grounding electrode, etc.), this problem can appear in a more severe form. This especially can be the case of HV distribution substations situated at the centers of consumption, i.e., deep within the city area, where normally there is insufficient space available for laying large grounding electrodes (grids) and where there are many public buildings in the vicinity [6]. Also at present, in the middle voltage networks, cables having an insulated outer sheath are mainly used, instead of the oil-filled cables whose contribution to the solution of the grounding problems of HV substations has been considered essential for a long time. Under such conditions, the grounding electrode (grid) potential rise can reach more than 1 kV and its reduction appears as a mandatory demand for substation designers. In such cases it can be shown that an efficient solution is to take measures that reduce the part of the ground fault current dissipated into the surrounding earth and that lead to an overall reduction of too high potentials appearing within the station and/or in its vicinity. This solution can be achieved by requiring a nonstandard HV cable construction of sufficiently low reduction factor value. However, especially if the feeding line is relatively long, this solution may prove very expensive.

The problem can also be overcome by installing a bare copper wire in the same trench and on the certain length *along* the feeding cable line [9]. Certainly, this wire will additionally reduce the current dissipated through the substation grounding system by providing an additional return path for the ground fault current. This way of solving the problem is more cost effective even in cases where there is enough free space available for laying a suficiently large ground grid. There are two essential reasons for this. First of all a strong inductive coupling of the buried copper wire with the cable line conductors significantly intensifies the fault current dissipation into the ground through the wire. Second, there is no need to dig a separate trench for the wire. Finally, this solution opens the possibility of using cheaper HV cables of standard constructive characteristics, even in extremely unfavorable grounding conditions.

As an illustration, this measure for obtaining a more favorable ground fault current distribution is symbolically depicted by Fig. 4.16.

As in the case in Section 4.3.1, the task is to determine the part of the ground fault current appearing at substation *B* due to the inductive coupling between the line conductors and the buried bare copper wire. This fraction of the ground fault current is diverted from the substation ground grid and at the point of discontinuity (end of the buried copper wire), *D* continues its flow toward its sources through the ground. With the aim to determine the total current diverted by the copper wire it is also necessary to determine the part of the ground fault current appearing in station *B* due the conductive coupling of the copper wire and the substation grounding electrode usually constituted of substation building foundation.

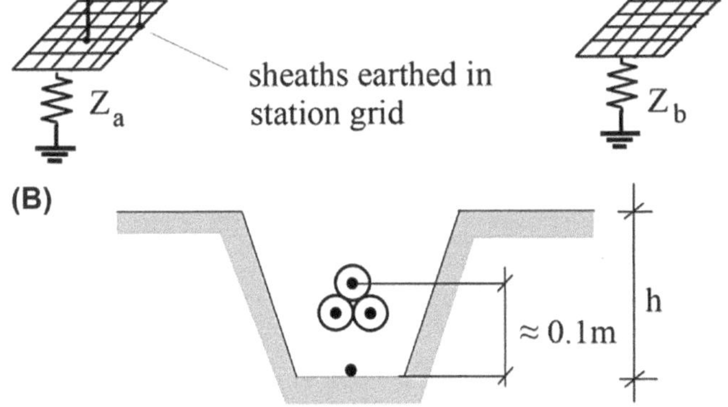

Figure 4.16 A nonhomogeneous cable line. (A) Cable feeding line. (B) Cables and copper wire [9].

4.3.2.2 The Equivalent Circuit

Let us assume that substation A is either source or transit and that substation B is supplied by a cable line, as shown in Fig. 4.16. Also let us assume that both of these substations belong to an HV network with directly grounded neutral points and that the current appearing in the foreseen copper wire has a negligible influence upon the value of the total ground fault current, which is a quit realistic assumption,

Then, by using as the theoretical foundations:

- the method of symmetrical components,
- the decoupling technique,
- the reduction of uniform ladder circuits to the one equivalent π - network (Eqs. 2.63 and 2.64), and
- the introduction of an equivalent metal sheath substituting all three actually existing cable line metal sheaths (Section 3.2).

The grounding system shown in Fig. 4.16 can be presented by the relatively simple equivalent circuit in Fig. 4.17.

The used notation has the following meaning:

U_1—voltage induced by current I_f in the equivalent cable line sheath,
U_2—voltage induced by current I_i in the equivalent cable line sheath,
U_3—voltage induced by current I_{she} in the copper wire,
U_4—voltage induced by current I_f in the copper wire,
I_f—ground fault current component passing through the grounded neutral points in the supply network,
I_{she}—fault current through the equivalent cable sheath only as a consequence of conductive connections with ground grids at its ends,

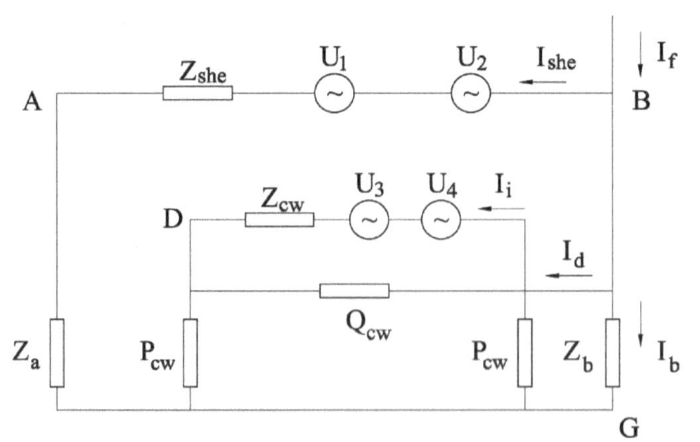

Figure 4.17 The equivalent circuit of the grounding system of Fig. 4.16.

I_b—fault current dissipating into the earth through the substation grounding system excluding the buried ground wire,
I_d—fault current diverted from the substation grounding system by the buried copper wire,
I_i—current source that replaces the influence of the inductive coupling between the cable line and buried copper wire,
Z_{she}—self-impedance of the equivalent cable line sheath,
Z_{cw}—self-impedance of the buried copper wire,
Q_{cw} and P_{cw}—impedances of the equivalent π-cell for the lumped parameter ladder circuit representing the buried copper wire as grounding conductor,
$Z_a(Z_b)$—grounding system impedance of substation A (B), and
G—remote ground.

For the known relevant parameters of the buried copper wire, including its length and the equivalent resistivity of the surrounding soil, impedances Z_{cw}, Q_{cw}, and P_{cw} can be determined by using Eqs. (2.63), (2.64), and (2.55), respectively. Also for the known value of the reduction factor of the cable line (complex number) the relevant parameters of the equivalent cable metal sheath can be determined by using Eq. (3.39) represented as a system of two equations. On the basis of these parameters and the certain value of the cable line length it is possible to determine impedance Z_{she}, all induced voltages U_1, U_2, U_3, and U_4, and all impedances Z_a, Z_b, Z_{she}, and Z_{cw} in the equivalent circuit of Fig. 4.17. Then, desired current I_d can be also determined. However, in many practical cases, impedance Z_a in comparison with impedance Z_{she} can be disregarded ($Z_a \approx 0$). Also the full effect of this measure is obtained when the length of the copper wire is so large that impedance Q_{cw} can be considered infinitely large. By using the equivalent circuit and these approximations, we can write the following system of equations:

$$\begin{aligned} Z_{she}I_{she} + Z_{m1}I_f + Z_{m2}I_i &= 0 \\ (Z_{cw} + P_{cw})I_i + Z_{m3}I_f + Z_{m4}I_{she} &= 0 \end{aligned} \tag{4.34}$$

where Z_{m1} is the mutual impedance between the faulted phase conductor and the equivalent cable line sheath, Z_{m2} is the mutual impedance between the equivalent cable sheath and the buried copper wire, Z_{m3} is the mutual impedance between the faulted phase conductor and the buried copper wire, and Z_{m4} is the mutual impedance between the equivalent cable line sheath and the buried copper wire.

Bearing in mind the imagined physical appearance and disposition of the equivalent sheath (conductor) one can see that impedances Z_{m2}, Z_{m3}, and Z_{m4} are mutually equal ($Z_{m2} = Z_{m3} = Z_{m4}$) and can be commonly denoted as Z_m. On the basis of this observation and by solving Eq. (4.34) we obtain:

$$I_{she} = \frac{Z_m^2 - (Z_{cw} + P_{cw})Z_{m1}}{(Z_{cw} + P_{cw})Z_{she} - Z_m^2} I_f \tag{4.35}$$

$$I_i = \frac{Z_m Z_{m1} - Z_{she} Z_m}{(Z_{cw} + P_{cw})Z_{she} - Z_m^2} I_f. \tag{4.36}$$

On the basis of Eqs. (4.35) and (4.36) the total reduction factor of the considered feeding line is:

$$r_t = \frac{I_f - I_{she} - I_i}{I_f} \approx \frac{(Z_{cw} + P_{cw} - Z_m)(Z_{she} - Z_{m1})}{(Z_{cw} + P_{cw})Z_{she} - Z_m^2} \tag{4.37}$$

Then, under the conservative assumption that the impedance Q_{cw} is infinitely large, the reduction of current I_b appearing only as a consequence of the conductive coupling between the copper wire and the substation grounding electrode, according to Fig. 4.16, can be expressed by:

$$r_c = \frac{P_{cw} Z_{she}}{P_{cw} Z_{she} + Z_b(P_{cw} + Z_{she})} \tag{4.38}$$

Finally, the current passing through the grounding system of supplied substation B is:

$$I_b = r_t r_c I_f = \frac{r_t P_{cw} Z_{she}}{P_{cw} Z_{she} + Z_b(P_{cw} + Z_{she})} I_f \tag{4.39}$$

In the case of application when the copper wire is laid along the entire line length Eq. (4.37) should be modified. Then, instead of P_{cw} should be Z_a if substation A is transit one, or should be omitted ($P_{cw} = 0$) if substation A is source one. Since the dimensions of the grounding electrode of the considered type of HV substations are relatively small, the proximity effect between the substation grounding electrode and buried copper wire is negligable [17]. However, this effect seen in principle can be approximately taken into account through the value of impedance P_{cc} by using Eq. (6.21).

Here it could also be interesting to see the directions of the currents induced in the copper wire on the lefft and on the right from the fault point, I_{il} and I_{ir}, when the fault occurs in the feeding cable line itself and in the vicinity of the supplied substation. For this purpose, under the previously adopted assumptions, we will use the circuit shown in Fig. 4.18.

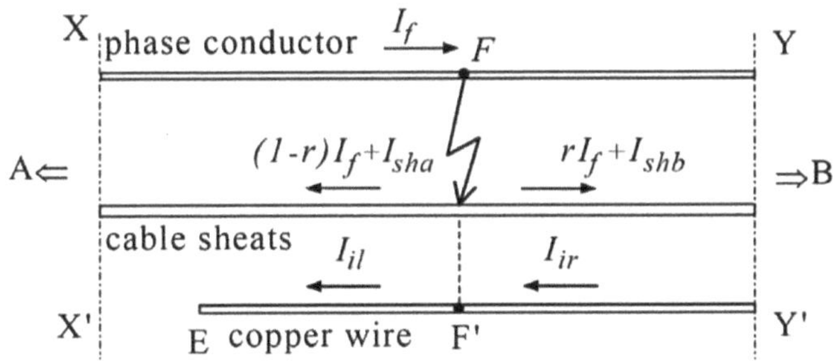

Figure 4.18 Current I_f distribution for a fault at an arbitrary point near substation *B*.

The notation used has the following meaning:

X (X'') and Y (Y')—end point of an arbitrary line segment, involving the end point of the cooper wire and the fault place,

E—end point of the buried copper wire,

I_{sha}(I_{sha})—current appearing in the metal sheath of the faulty cable as a consequence of its conductive connection to the ground grid of substation A (B).

To provide a clearer insight, the given circuit does not show the component of the ground fault current returning through the grounded neutral(s) in substation *B*. However, this current can be disregarded, because for faults near substation B this current almost completely returns into the HV network via the cable sheath and grounding connections in this substation.

As can be seen, the current circulating through the metal sheath of the faulty cable toward station A is enlarged by current I_{sha}, while the current circulating toward substation B is enlarged by current I_{shb}. As a consequence, the current through the substation B ground grid is apparently reduced, as shown in Section 3.3.

4.3.3 Quantitative Analysis

4.3.3.1 Overhead Feeding Line

This analysis is performed with the aim of illustrating the efficiency of the described measures for reducing the fault current dissipated through the ground grid of a supplied HV substation.

It is assumed that substation B in Fig. 4.1 is supplied by a 110 kV line equipped with an ACSR 95/55 mm^2 ground wire. The line can be treated as infinitely long, thus Z_d is equal to $Z_{gw\infty}$, or $(1.253 + j0.846)\ \Omega$. This value is used in all calculations presented here, independently of the adopted value of equivalent soil resistivity, ρ. Also it is assumed that the line from point D toward station A is transposed. The tower's height is 23.4 m and its configuration is shown in Fig. 4.12A.

The counterpoise conductor is a copper wire of cross-section 120 mm^2 laid along two nearest spans of an overall length of 500 m and buried to a depth of 0.8 m. The ground grid is at the same depth and has dimensions 80 m × 100 m and spacing of 10 m. Calculations of the achieved grade of the ground grid current reduction have been performed for the following soil resistivity: 20, 50, 100, and 200 Ωm, covering the relatively wide scale of the practically possible values.

On the basis of the here presented analytical expressions and the given data, it is obtained that the ground grid current is 31%, 25%, 20%, and 17% of I_f for $\rho = 20$, 50, 100, and 200 Ωm, respectively.

The fault current diverted from the ground grid and injected into the earth by the ground wire and the counterpoise conductor is 37%, 41%, 44%, and 46% of I_f, for $\rho = 20$, 50, 100, and 200 Ωm, respectively.

The grid current without application of the counterpoise would be 64%, 57%, 48%, and 37% of I_f, for $\rho = 20$, 50, 100, and 200 Ωm, respectively.

If we bear in mind that the necessary length of the copper wire for the considered ground grid is 1780 m, according to the presented results it is clear that the efficiency of the ground fault current dissipation by the counterpoise conductor is significantly higher. Of course, this can be explained by the fact that this conductor acts as an external electrode distant from other elements of the ground grid. Besides that, the additional currents are absorbed by this conductor because of its inductive coupling with the feeding line and the conductive connections with the feeding line ground wire. It is especially important to note that the effect of reduction is more pronounced under less favorable natural conditions (higher ρ), i.e., when it is more needed.

The obtained results point to the following conclusion: in many cases length of the ground grid conductor can be restricted only to the required connections between the metal structures to be earthed because it is more efficient/economical to invest in the counterpoise conductor than to invest further in the substation ground grid.

4.3.3.2 Cable Feeding Line

Let us assume that a substation will be supplied by a 110 kV cable line 3 km long. The foreseen cable type is XHE 94-A. In the standard design its sheath has the other diameter equal to 91 mm, and is made of copper strings with cross-section equal to 95 mm^2. By using Eq. (3.16) the reduction factor of such cable line when its cables are laid in three-angular formation is 0.020 − j0.106. By selecting the next largest sheath cross-section offered by the manufacturer's catalog equal to 120 mm^2, the value of the cable line reduction factor would be 0.015 − j0.084. If a higher reduction of current I_b is necessary, the next largest cross-section is 185 mm^2 and then the reduction factor of the considered cable line would be 0.008 − j0.055.

The described measurement of reducing earth current I_b will be considered under the assumption that the copper wire has a cross-section of 120 mm^2 and is buried to a depth of 1 m together with the cables (Fig. 4.16). Also it will be assumed that the equivalent soil resistivity is relatively high or $\rho = 200$ Ωm, and according to this the effective value of grounding impedance of the supplied substation $|Z_b|$ is relatively high and is equal to 0. 4 Ω.

We will consider the reduction of current I_b for the different lengths of the copper wire. Thus we will assume that the length of the copper wire, L_{cw}, one after another, takes the following values: 500, 1000, and 1500 m. On the basis of these data and by using Eqs. (4.37) and (4.38), we obtain that the achieved reductions are:

- $r_t = 0.0403 - j0.0956$ and $r_c r_t = 0.0386 - j0.0611$, for $L_{cw} = 500$ m and
- $r_t = 0.0149 - j0.0949$ and $r_c r_t = 0.0155 - j0.0318$, for $L_{cw} = 1000$ m.

Under conditions where the substation grounding electroge has already been extended or reinforced through the use of vertical earth rods, satellite earth grids, etc., reduction of ground grid current I_g becomes the only solution. Then, the designers, generally speaking, have two possibilities:

- Application of the measure considered herein, or
- Placing an order with the manufacturers to fabricate a specially designed cable of sufficiently low reduction factor value.

Of course, the following question is raised: which of these two possible solutions is more economical? On the basis of the given data it is not difficult to show that the reduction of 24.36% and 50.7% obtained by enlarging the cable sheath cross-section requires the additional consumption of copper strings that are 1990 and 7168 kg, respectively. However, for reductions of 29.3%, and 46.2% obtained by the buried copper wire only 531 and 1062 kg, are necessary.

Based solely on the presented results, it is obvious that the advantage is on the side of the solution described herein. Certainly, this general conclusion does not exclude separate considerations in certain concrete cases. The need for the considered measure can be expected mainly in the case of HV substations located deep within the city area and surrounded with relatively old buildings without foundation earth electrodes.

REFERENCES

[1] B. Thapar, S. Madan, Current for design of grounding systems, IEEE Transactions on Power Apparatus and Systems 100 (7) (1981) 342–345.
[2] Lj.M. Popović, Determination of the reduction factor for feeding cable lines consisting of three single-core cables, IEEE Transactions on Power Delivery 18 (3) (2003) 736–743.
[3] Lj.M. Popović, Improved analytical expressions for the determination of the reduction factor of the feeding line consisting of tree single-core cables, ETEP European Transactions on Electrical Power 18 (2) (2008) 190–203.
[4] Lj.M. Popović, Influence of the metallic installations surrounding the feeding cable line on the ground fault current distribution in supplied substations, IEEE Transactions on Power Delivery 23 (4) (2008) 2583–2590.
[5] Lj.M. Popović, Ground fault current distribution when a ground fault occurs in HV substations located in an urban areas, Progress in Electromagnetic Research B 59 (2014) 167–179.
[6] J.E.T. Villas, D. Mukhedkar, V.R. Fernandesm, A.C. Magalhaes, Ground grid design of a transition station system – a typical example of fault transfer, IEEE Transactions on Power Delivey 5 (1) (1989) 124–129.
[7] S.T. Sobral, V.S. Costa, M.S. Campas, D. Mukhedkar, dimensioning of nearby substations interconnected grounding system, IEEE Transactions on Power Delivery 3 (4) (1988) 1605–1614.
[8] Lj.M. Popović, Practical method for evaluating ground fault current distribution in station supplied by an unhomogeneous line, IEEE Transactions on Power Delivery 12 (2) (1997) 722–727.
[9] Lj.M. Popović, Efficient reduction of fault current through the grounding grid of substation supplied by cable line, IEEE Transactions on Power Delivery 15 (2) (2000) 556–561.
[10] Lj.M. Popović, Efficient reduction of fault current through the grounding grid of a substation supplied by an overhead line, ETEP European Transactions on Electrical Power 16 (3) (2006) 269–274.
[11] Lj.M. Popović, Ground fault current distribution in a station supplied by a line composed of a cable an overhead section, ETEP European Transactions on Electrical Power 17 (3) (2007) 207–218.
[12] S. Mangione, A simple method for evaluating ground-fault current transfer at the transient station of a combined overhead cable line, IEEE Transactions on Power Delivery 23 (3) (2008) 1413–1418.
[13] G. Zizzo, A. Campoccia, E. Riva Sanseverino, A new model for safety analysis in a system of earthing systems interconnected through bare-buried conductors, COMPEL Computation on Mathematics in Electrical Engineering 28 (2009) 412–436.

[14] Short-circuits currents in three-phase A.C. system-Part 3: currents during two separate simultaneous line-to-earth short circuit and partial short-circuit flowing trough earth. IEC Int. Std. 60909−3, Ed. 3.0, 2009-03.
[15] IEEE Guide for Safety in Substations Grounding, Std. 80.
[16] Lj.M. Popović, Active length of cable lines with uncoated metal sheaths as long grounding conductors, ETEP European Transaction on Electrical Power 22 (2) (2012) 206−215.
[17] Lj.M. Popović, Proximity effect between an earthing grid and external electrodes in an earthing system of high-voltage installations, IEE Proceedings − C 133 (6) (1986) 346−352.

CHAPTER 5

Critical Ground Fault Position

5.1 INTRODUCTION

A ground fault at any place within a high-voltage (HV) power network causes large currents accompanied by erased potentials at different places within this network. The magnitude of these currents and potentials is not the same for different places of faults within the same HV network. Thus to securely and economically provide protection against undesired consequences, such as loss of human life, burning down of transmission ground wires, and damage to sensitive telecommunication equipment in HV substations, it is necessary to evaluate, as precisely as possible, the ground fault current and its distribution under the worst fault position. However, in certain cases the critical fault position is not easily predictable because the corresponding ground fault current distribution cannot be estimated by using relatively simple calculations. In the case of a generating HV substation, prediction of the maximum current flowing between grounding system and the surrounding earth cannot be made without analysis of numerous ground faults in the outgoing overhead lines [1—4].

All potentials and potential differences relevant for estimating the achieved safety conditions (touch and step voltages) are proportional to the current dissipated through the substation grounding system into the surrounding earth, later on simply called earth current. Since safety conditions should be satisfied under all practically possible fault conditions, it is necessary to predict the value of this current under conditions when it reaches its largest value. The problem becomes complex when it is not obvious that the place of the worst (critical) fault is within the considered substation itself, i.e., when to obtain the largest value of earth current, ground fault places outside of the considered HV substation have to be explored. Such a situation appears in the case of HV source stations supplying large urban agglomerations, as well as in the case of large power stations. For such substations, it is necessary to perform a large number of consecutive ground fault analyses assuming the fault place at many towers of one or more outgoing overhead lines [5—8]. In these cases, one portion of ground fault current that returns via the earth into the power system

Practical Methods for Analysis and Design of HV Installation Grounding Systems
ISBN 978-0-12-814460-2
https://doi.org/10.1016/B978-0-12-814460-2.00005-2

must pass through the generating substation grounding system to the grounded transformer neutral(s). Thus this ground fault current portion produces a substation ground grid rise that should be determined for different possible ground fault locations.

In modern electric power engineering practice, determination of ground fault current distribution at different places is important for two additional reasons. The first is selection of the size of the transmission line ground wire capable of withstanding thermal stress during ground fault [6,8,9], and the other is estimation of the potential rise of the faulty tower [7,10].

Extensive research has been undertaken, especially during the 1980s and 1990s, to model transmission networks for the purpose of ground fault current analysis. The advantages and drawbacks of a great number of developed methods and procedures are presented in detail in Ref. [5]. Thus here it will be pointed out that in their development and improvement only two tendencies can be clearly distinguished. On the one hand, an effort is made to make these methods more convenient in applications [3,4,8,11,12] due to the large number of cases that have to be analyzed. On the other hand, the authors try to improve further the methods by including new, less important factors. In some especially complex and specific conditions they could contribute to obtaining a technically and economically acceptable solution [1,2,4]. However, here it is important to bear in mind that the nature of the considered problem is such that only approximate accuracy is possible. Some of the input data (e.g., value of a tower footing resistance) cannot be exactly known at the design stage of certain generating substations and their outgoing lines. Thus the methods of high accuracy allowing appreciation of many relevant factors and parameters, including those of lower significance, cannot be considered as preferable because at the design stage of an HV generating substation its high accuracy cannot be practically utilized. Independently of that, more accurate methods have been developed by using special matrix techniques and modern computers. However, experience gained in their application has shown that relative error propagation can be a problem when solving a large complex matrix.

A way of overcoming the problem has been attempted by finding a solution for lumped parameter ladder circuits. Since the late 1960s many authors (Endrenyi, Poter, Finsh, Mukhedkar, Sebo, Dawalibi, etc.) have presented methods of obtaining the solution to these circuits. The generally used approach was to represent the lumped parameter ladder circuit by an equivalent distributed parameter ladder circuit and to solve it by differential equations. However, these methods give satisfactory results only when the line or its sections considered in this manner has a sufficient number of spans.

Then, by using Kirchhoff's rules, the principle of superposition, and the summation of different geometric series, equations were derived that took into consideration the discrete nature of lumped parameter ladder circuits [10,11]. However, the procedure was not finalized, thus the solution was valid only for specific cases from the point of view of possible terminal conditions. Finally, by developing the general equations of the line represented by discrete parameters, the problem of lumped parameter ladder circuits was treated as definitively solved [12]. However, by introducing a certain correction, the exact and complete deriving procedure of these equations is presented just in Section 2.1. Fortunately, the omission made in this paper had no influence on the somewhat later method developed for determining the critical ground fault position [1].

By applying this solution and the previously known analytical procedures (the method of symmetric components, driving point technique, and decoupling technique) a practical method of sufficiently high accuracy at the design stage has been developed [1]. In this way the considered problem has become solvable with no need to use a computer, even in cases where many fault places outside a generating substation need to be considered.

Somewhat later, the developed practical method was modified to include the specific case where ground fault analysis should be performed for a double-circuit parallel line [13]. The lines performed as a double-circuit line, in general, diverge at some point and terminate at different substations (sometimes at both ends). However, when these lines are parallel (the lines start and finish at the same substations by using the same towers), somewhat more intensive ground fault currents appear at fault places in comparison to single-circuit lines under identical all other relevant conditions. This is the reason why this specific type of overhead line deserves special consideration and development corresponding to analytical procedures necessary for determining the critical ground fault position.

5.2 GROUND FAULT IN OUTGOING TRANSMISSION LINES

5.2.1 Ground Fault at an Arbitrary Place

In designing grounding systems of HV substations, one of the objectives is to find an economical solution, which means that it should be in accordance

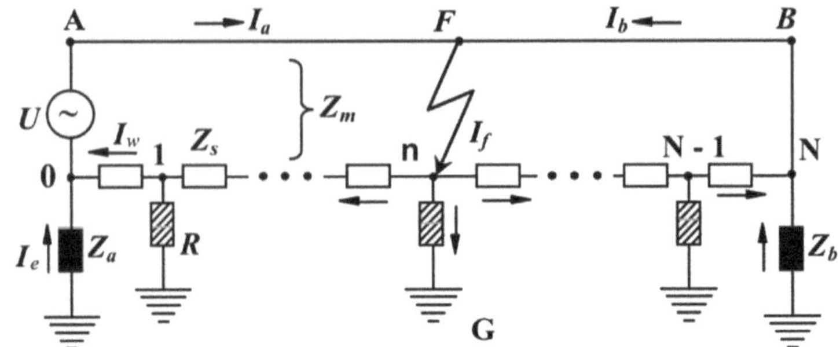

Figure 5.1 Parts of the ground fault current for a fault at an arbitrary line tower [1].

with real needs, i.e., without unnecessary expenditure. This can be done only if the maximal (critical) earth current is exactly determined. However, before doing this we have to know exactly the critical fault position. The problem appears in the case of HV substations where this place should not be obviously in the framework of these substations. This is the case with generating (power and source) stations. For these substations, as already mentioned, the number of ground fault studies that should be performed can be very large.

In principle, a ground fault anywhere in an HV network causes fault currents through the grounding systems of all substations with grounded neutral(s). However, to find the worst ground fault position for a certain generating substation, it is necessary to analyze faults occurring in both this substation and in its outgoing overhead lines. A complete analysis can show that the worst fault is placed in one of the outgoing lines at a certain critical distance from the considered substation.

The overhead line through which a source substation A supplies a distribution substation B, under conditions of a ground fault at an arbitrary point, F, along the line, is schematically presented in Fig. 5.1.

The notation used in this circuit has the following meaning:

U—source voltage equal to the phase voltage in station A,

I_f—total ground fault current,

$I_a(I_b)$—ground fault current component to the left (right) of the fault place, F,

I_e—earth current returning through the grounding system of substation A,

I_w—fault current returning through the ground wire(s) of the first span, counted from station A,

$Z_a(Z_b)$—grounding system impedance of substation A (B), which does not include the grounding effects of the ground wire(s) of the considered line,

Z_s—self-impedance of the ground wire(s) per span,

Z_m—mutual impedance between the ground wire(s) and the faulty phase conductor per span,

R—average tower footing resistance,

n—number of spans to the fault location, counted from station A, and

N—total number of the line spans.

The equivalent circuit of Fig. 5.1 shows parts of the fault current when a single line-to-ground fault occurs at an arbitrary (nth) tower along the line.

In forming the presented overhead line model, the following approximations and idealizations were used:

- Phase and ground wire(s) impedances and their mutual impedances are identical to the values calculated on the basis of infinite transmission lines.
- Tower footing resistances are mutually equal and any mutual interference to their own earth current is negligible.
- Impedances of the ground wire(s) between two adjacent towers are mutually equal.

As is well known in practice, values of the grounding resistances of towers are not mutually equal. However, at the design stage we cannot obtain exact data regarding any of the tower footing resistances. Therefore we are forced to adopt an approximately estimated value, which is then assigned to a larger number of line towers. This estimation is based on earth geological characteristics common to the sites where the towers will be constructed.

Impedances Z_s and Z_m are calculated by using formulae based on Carson's theory of the ground fault current return path. Impedance Z_m is determined under the realistic assumption that a line section of several spans from A to F is not transposed. Since current I_e is to be determined under the worst fault conditions, we will assume that the faulted conductor is the most distant from the ground wire(s).

Current I_a returns through the neutral(s) in station A, whereas current I_b returns through the neutral(s) in substation B. When the fault occurs inside station A, current I_a circulates only in the axial direction through the ground grid elements, thus it cannot create any potential toward remote earth irrespective of its value. However, when the fault occurs in the outgoing line, value of current I_a decreases in magnitude, but a part of it, current I_e (Fig. 5.1), returns through earth and creates a certain potential on the grounding system of substation A. The remaining part of current I_a, current I_w, returns to the power system through the ground wire(s) and grounding connections of substation A. The nearer fault produces the larger current I_a, but most of it, due to the proximity of substation A, returns through the ground wire(s). When the distance of the ground fault place from the considered substation increases, current I_e also increases. Thus it is necessary to examine at which distance current I_e reaches its maximal value.

By applying the:

- method of symmetric components,
- driving point technique,
- decoupling technique, and
- general solution of the lumped parameter ladder circuits in Section 2.1, the power system shown in Fig. 5.1 can be presented by the equivalent circuit in Fig. 5.2.

The used notation has the following meaning:

U_{ph}—auxiliary driving source at the fault place, whose voltage is the same as the phase voltage in supply station A,

V_n—potential of the faulted tower,

$I_{ia}(I_{ib})$—current source used to replace the influence of the inductive coupling between the ground wire(s) and the faulted phase conductor,

Z^+, Z^-—positive and negative sequence impedances of the electric power system seen at the fault place,

$Z_A^0\left(Z_B^0\right)$—zero sequence impedance of the electric power system at station A (B),

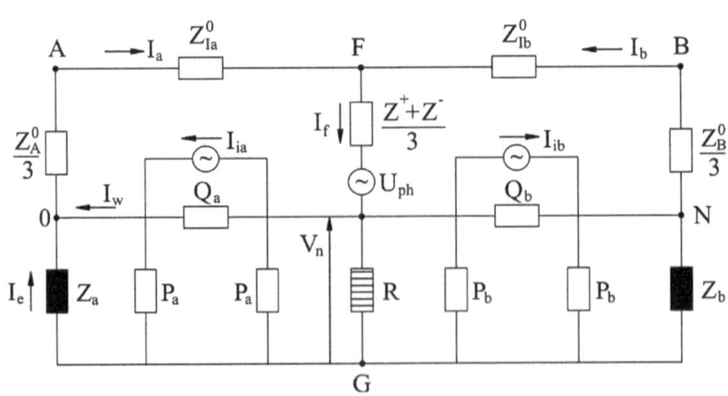

Figure 5.2 The equivalent circuit necessary for ground fault analysis [1].

$Z_{1a}(Z_{1b})$—zero sequence impedance of the line from the fault place to station A (B), and

G—remote ground.

Currents of the presented current sources are determined by:

$$I_{ia} = \frac{Z_m}{Z_s} I_a, \tag{5.1}$$

$$I_{ib} = \frac{Z_m}{Z_s} I_b. \tag{5.2}$$

Equivalent impedances Q_a and P_a, according to Eqs. (2.65) and (2.66), are determined by:

$$Q_a = \frac{k^n - k^{-n}}{1 + k} Z_\infty \tag{5.3}$$

$$P_a = \frac{k^n - k^{-n}}{k^n + k^{-n+1} - k - 1} Z_\infty, \tag{5.4}$$

where:

$$k = 1 + \frac{Z_\infty}{R}. \tag{5.5}$$

In expression (5.5), parameter Z_∞ designates the grounding impedance of the ground wire of an infinite line that is, in accordance with Eq. (2.4), determined by:

$$Z_\infty = \frac{Z_s}{2} + \sqrt{RZ_s + \frac{Z_s^2}{4}}. \tag{5.6}$$

Impedances Q_b and P_b can also be determined from expressions (5.3) and (5.4), but then it is necessary to introduce the corresponding number of spans, i.e., to replace n by $m = N - n$.

On the basis of the equivalent circuit of Fig. 5.2 and the given expressions, by varying n and by performing sequential calculations, it is possible to determine, for any line, the largest values of the current I_e, current I_w, and potential of any tower, V_n. Such analysis can be performed for a line of an arbitrarily complex power system, which means multiple lines, multiple power stations, multiple substations, different voltage levels, substations with neutral points either connected through low impedance to the ground or directly grounded, etc. Before this, it is necessary to calculate the positive, negative, and zero sequence impedances of the system at terminal points of the considered line. Also to calculate I_e, the equivalent circuit should be formed in such a way to present one branch that directly connects points A and G (Fig. 5.2). The equivalent impedance of this branch should be determined so that current I_a is divided into two constitutive components. One of them circulates through the neutral point(s) in station A, whereas the other circulates through the remaining grounded neutral points in the supply network, with the exception of the part of the network supplied by the considered line.

The solution of the lumped parameter ladder circuits presented in Chapter 2 can also be applied to reduce the amount of calculations required by the methods that, by treating each phase conductor separately, aim at either improving the accuracy or including in the analysis other types of faults.

5.2.2 Critical Fault Position at a Certain Distance Along the Line

In practice, the neutral points in an HV distribution substation are directly grounded; however, their contribution to the ground fault current in many cases is negligible $Z_A^0 \ll Z_B^0$. Thus the branch carrying current I_b in the equivalent circuit of Fig. 5.2 can be disregarded and, according to this, the current source I_{ib} as well. At the same time, this means that influence of the ground wire to the right of the fault location (Figs. 5.1 and 5.2) can be represented only by its grounding impedance, which is, according to Eq. (2.69), determined by:

$$Z_{nb} = \frac{(k^m + k^{-m+1}) Z_b Z_\infty + (k^m - k^{-m}) Z_\infty^2}{(k^m - k^{-m+2}) Z_b + (k^m + k^{-m+1}) Z_\infty}, \tag{5.7}$$

or if this impedance incorporates footing resistance of the faulted tower (Fig. 5.1), then:

$$Z'_{nb} = Z_{nb} - Z_s. \tag{5.8}$$

Also, in many practical cases, impedances Z_a and Z_b have such small values (for substations in 110 kV networks and over, often in the range from 0.05 to 0.1 Ω) that their influence on the critical value of I_e can be neglected. In this way we eliminate the necessity of preliminary determination of the grounding impedances of substations at the line ends, while the corresponding equivalent circuit and the whole calculation procedure become significantly simpler.

Impedance Z_f on the circuit in Fig. 5.4 is determined by:

$$Z_f = Z_A + nZ_s, \tag{5.9}$$

where:

$$Z_A = \frac{Z_A^+ + Z_A^- + Z_A^0}{3}, \tag{5.10}$$

$$Z_s = \frac{2Z_{ls}^+ + Z_{ls}^0}{3}, \tag{5.11}$$

whereas Z_A^+, Z_A^- and Z_A^0 represent the positive, negative, and zero sequence system impedances at substation A.

Impedances Z_{ls}^+ and Z_{ls}^0 in Eq. (5.11) represent the positive and zero sequence impedances of the line per one span. Impedances Z_l^+ and Z_l^0 of the line per unit length are, in accordance with Refs. [13,15], determined by:

$$Z_l^+ = R'_{ph} + j\omega\frac{\mu_0}{2\pi}\left(\frac{1}{4} + \ln\frac{D_{ph}}{r_{ph}}\right), \ \Omega/\text{km}, \tag{5.12}$$

$$Z_l^0 = Z'_{ph} - \frac{Z'^{\,2}_m}{Z'_{gw}}, \ \Omega/\text{km}, \tag{5.13}$$

where Z'_{ph} is the self-impedance of the phase conductor (Ω/km), $Z'_m$is the mutual impedance between phase conductors and the ground wire (Ω/km), $Z'_{gw}$is the self-impedance of the ground wire(s) (Ω/km), R_{ph} is the phase conductor resistance (Ω/km), D_{ph} is the geometric mean distance between the phase conductors (m), and r_{ph} is the geometric mean radius of the phase conductor (m).

On the basis of the previous considerations, the circuit of Fig. 5.1 can be substituted by the simpler circuit in Fig. 5.3, whereas the equivalent circuit of Fig. 5.2 can be substituted by the simpler circuit in Fig. 5.4.

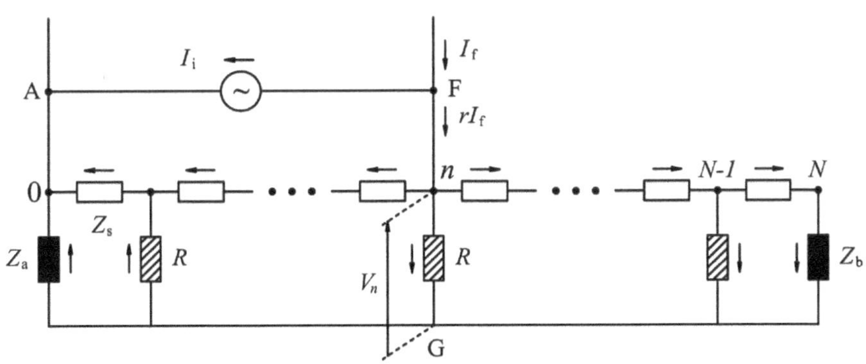

Figure 5.3 The return paths of the ground fault current [14].

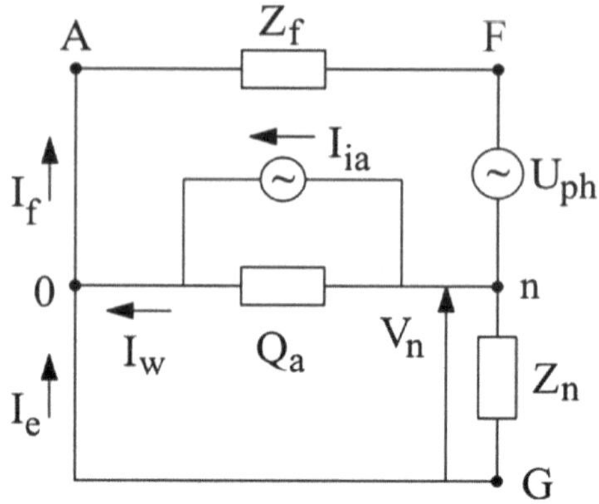

Figure 5.4 Simplified equivalent circuit [1].

According to the given equivalent circuits (Figs. 5.2 and 5.4), impedance Z_n represents the grounding impedance at the fault place determined by:

$$Z_n = \frac{Z'_{nb}(Z_b = 0)P_a}{Z'_{nb}(Z_b = 0) + P_a}, \tag{5.14}$$

where impedance $Z'_{nb}(Z_b = 0)$, from expressions (5.7) and (5.8), is:

$$Z'_{nb}(Z_b = 0) = \frac{k^m - k^{-m+1}}{k^m + k^{-m+1}} Z_\infty - Z_s. \tag{5.15}$$

In cases where the line can be considered infinite, the following, still simpler, expression can be used:

$$Z'_{nb}(\infty) = Z_\infty - Z_s = \sqrt{RZ_s + \frac{Z_s^2}{4}} - \frac{Z_s}{2}. \tag{5.16}$$

Certainly, the same equivalent circuit can be used for estimating the maximum value of current through the transmission line ground wire, I_w. In that case the worst position of the fault is obvious, i.e., a priori known. This is the tower closest to supply substation A, so that m is equal to $N - 1$ ($m = N - 1$) and we have:

$$Z_{1b}(Z_b = 0) = \frac{k^{N-1} - k^{-N}}{k^{N-1} + k^{-N}} Z_\infty. \tag{5.17}$$

Then, the equivalent circuit in Fig. 5.4 can also be used for estimation of the potential of the faulted tower, V_n. When the line is long enough and the place of the fault sufficiently far away from the line ends so that the sections left and right of the fault place can be considered infinite, impedances Q_a and Q_b become infinitely large and the branches with these impedances (Fig. 5.2) can be omitted, while the expression for impedance Z_n, taking into consideration expressions (5.6) and (5.16), reduces to the following very simple form:

$$Z_n = \frac{Z_{na}(\infty) Z'_{nb}(\infty)}{Z_{na}(\infty) + Z'_{nb}(\infty)} = \frac{\sqrt{RZ_s}}{2}. \tag{5.18}$$

A line, or its section containing N spans, can be considered infinite from the point of view of the grounding effects if, according to Eq. (2.80), the following condition is satisfied:

$$N_a(\varepsilon = 0.05) \approx \frac{3}{\ln\left(1 + \sqrt{\frac{|Z_s|}{R}}\right)}, \tag{5.19}$$

where N_a is the active line length expressed through the number of average spans and ε is the desired relative accuracy of the grounding impedance of the ground wire(s) of length equal to its active length (an arbitrarily small, a priori adopted, number).

The results of the quantitative analysis of the grounding impedance, Z_n, variations at the fault place along different transmission lines are shown in Figs. 5.7 and 5.8 (Section 5.2.4). On the basis of the shown diagrams it is not difficult to see that this impedance is dependent on the fault place. As a result, it has a certain influence on the value of the total ground fault current and its distribution along the line. The results of the quantitative analysis of this influence are presented in Fig. 5.9. They show that, depending on the line length, the worst fault position is at the end of the line or at a certain critical distance from the line beginning, which we shall designate as c. Thus, with the aim of simplifying the problem, for the beginning we shall assume that the considered line is infinitely long. In this way we can be sure that the critical position of the fault cannot be at the end of the line, but only somewhere along the line.

On the basis of previous assumptions the position of the fault that is critical from the standpoint of the value of I_e can be determined by using the following well-known mathematical condition: the first derivative of the function in terms of the independent variable must be equal to zero, i.e., the small variation of the considered function, $I_e(n)$, near its extreme value must be negligible. This criterion in the considered case can be expressed, as seen from the circuit in Fig. 5.4, by the following approximation:

$$\frac{Q_a(c)I_f(c)}{Q_a(c)+Z_n(c)} \approx \frac{Q_a(c+1)I_f(c+1)}{Q_a(c+1)+Z_a(c+1)}. \tag{5.20}$$

By setting Eq. (4.20) algebraically one obtains:

$$\frac{Z_n(c)}{Q_a(c)I_f(c)} - \frac{Z_n(c+1)}{Q_a(c+1)I_f(c+1)} \approx \frac{1}{I_f(c+1)} - \frac{1}{I_f(c)}. \tag{5.21}$$

The value of impedance Z_n at the worst/critical fault position, c, and its value at the towers immediately adjacent to c, in practical conditions, differ very little, so that we can write:

$$Z_n(c) \approx Z_n(c+1). \tag{5.22}$$

On the basis of this approximation and the equivalent circuit of Fig. 5.4, it follows that:

$$\frac{1}{I_f(c)} - \frac{1}{I_f(c+1)} \approx \frac{Z_s}{U_{ph}}. \tag{5.23}$$

By using approximations (5.22) and (5.23), expression (5.21) can be transformed into:

$$\frac{Q_a(c+1)}{Q_a(c)}\frac{1}{I_f(c)} - \frac{1}{I_f(c+1)} \approx \frac{Z_s}{U_{ph}}\frac{Q_a(c+1)}{Z_n(c)}. \tag{5.24}$$

With reference to the results presented in Fig. 5.9, distance c can be considered so large that one more approximation can be introduced:

$$\frac{Q_a(c+1)}{Q_a(c)} \approx k. \tag{5.25}$$

Now, under the previously made assumption that the line is infinitely long, according to the results of the presented quantitative analysis (Section 5.2.4), the critical fault position in the line remains unchanged. Then, by using Eq. (5.18) it is possible to write:

$$Z_n(c) \approx \frac{k^c+1}{2k^c}Z_\infty \approx \frac{Z_\infty}{2}. \tag{5.26}$$

Finally, from Eqs. (5.26) and (5.3) we obtain:

$$\frac{Q_a(c+1)}{Z_n(c)} \approx \frac{2}{k+1}(k^c-1). \tag{5.27}$$

By introducing approximations (5.20)−(5.22) and by using the equivalent circuit of Fig. 5.4, expression (5.27) can be transformed into the following equation:

$$c + \frac{2Z_A + (1-k)Z_s + rZ_\infty}{2Z_s} = \frac{2}{k^2 - 1}k^{c+1}, \tag{5.28}$$

where coefficient r represents the reduction factor of the considered overhead line that is defined by the following well-known expression:

$$r = 1 - \frac{Z_m}{Z_s}. \tag{5.29}$$

Since Eq. (5.28) represents a transcendental equation, its solution is possible only graphically. As an illustration, for the numerical example given in the quantitative analysis presented in this section, aluminum conductor steel-reinforced (ACSR) ground wire and $R = 30\ \Omega$, we obtain that the critical position is 17.2 ($c = 17.2$), as shown in Fig. 5.5.

The used notation has the following meaning:

F_1 and F_2—right and left side of Eq. (5.28) as a function of the fault place, n, encountered from substation A (Fig. 5.1).

Eq. (5.28) is not quite accurate, but it significantly reduces the necessary sequential (from tower to tower) ground fault analyses. Certainly, since in practice c can only be an integer, the final solution must

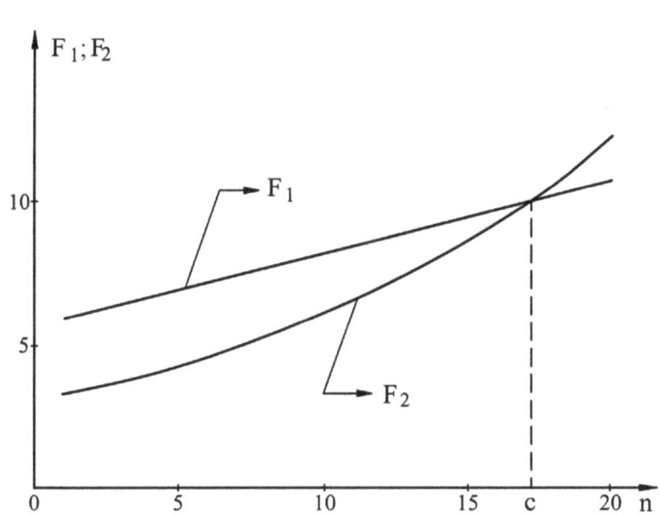

Figure 5.5 An illustration solving Eq. (5.28).

be sought among the nearest integers. Making use of the circuit shown in Fig. 5.4 the sequential analyses for $n = 17-19$ show that the critical fault position is the 18th tower.

5.2.3 Critical Fault Position at the End of the Line

The fault position determined in the previously described procedure and the corresponding value of current I_e can be considered critical only if the line is sufficiently long. According to the consideration given in Section 5.2.4, it is necessary that the total number of spans is greater than the limiting number of spans N_l. If the condition is not satisfied ($N < N_l$), the worst position of the fault is at the line end. By using the equivalent circuit of Fig. 5.4, in the case $n = N$ the corresponding earth current, under the assumption $Z_b = 0$, is:

$$I_e(N) = rI_f(N) = \frac{rU_{ph}}{Z_A + NZ_s}. \tag{5.30}$$

The limiting length can be determined from the condition of equality of current I_e when the fault is at the end of the line and when, under the assumption of an infinite line, the fault is at critical distance c. This condition, according to Eq. (5.30), can be expressed by:

$$\frac{rU_{ph}}{Z_A + N_l Z_s'} = I_e(c). \tag{5.31}$$

From Eq. (5.31) it follows that:

$$N_l = \mathrm{Re}\left(\frac{rU_{ph} - Z_A I_{emaxc}}{Z_s' I_{emaxc}}\right), \tag{5.32}$$

where Re is the real part of the expression in parenthesis and $I_e(c)$ is the maximum value of current I_e obtained under the assumption of an infinitely long line.

As can be seen from the outlined procedure, the critical value of current I_e for a line of arbitrary length can be determined through a reduced calculation procedure. When the considered generating station has many outgoing lines, the numerical method, i.e., the sequential (from tower to tower) ground fault analyses, seen in principle, should be applied to each of them separately. Therefore by applying the presented analytical method the large number of ground fault analyses can be significantly reduced so that for solving the considered problem a computer application is not required.

5.2.4 Quantitative Analysis

Quantitative analysis has been performed with the aim of giving an insight into the influence of the relevant factors and parameters upon the ground fault current distribution and the critical ground fault position. With this aim, we have considered the influences of the following factors: total line length, type of material used for the ground wire, and the tower footing resistance.

Let us assume that the line, which connects substations A and B in Fig. 5.1, is a 110 kV line with aluminum/steel 240/40 mm^2 phase conductors. Also let us assume that the line can be equipped with different ground wires. Thus we will assume steel 50 mm^2 in one case and ACSR 95/55 mm^2 in the other.

Line impedances per one span are determined on the basis of the following assumptions: average length of the span is 250 m and equivalent soil resistance along and around the line is 50 Ωm. The corresponding resistances per unit length of the phase conductors, steel and ACSR ground wire, are: 0.12, 3.5, and 0.299 Ω/km, respectively, while geometric mean radii are 8.53, 0.00249, and 6.23 mm, also respectively. The line configuration in all the considered cases is shown in Fig. 5.6.

On the basis of the given data, impedance Z_n as a function of the fault location along the line and for different line lengths, ground wires, and average tower footing resistances is graphically presented in Figs. 5.7 and 5.8.

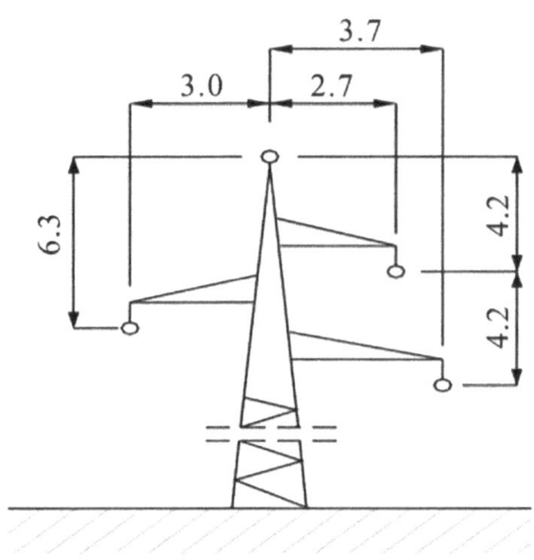

Figure 5.6 Disposition of line conductors (dimensions in meters) [1].

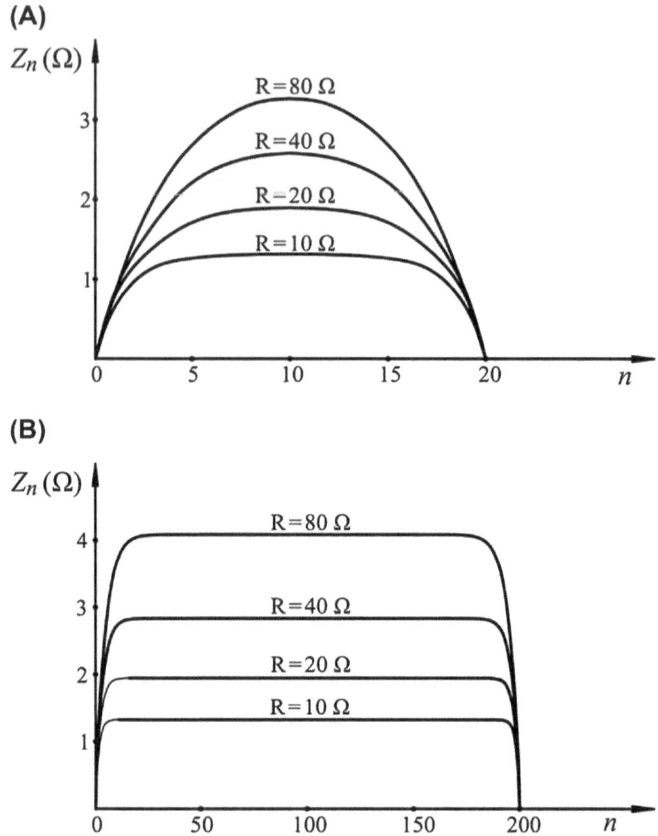

Figure 5.7 Impedance Z_n in the case of a line equipped with the steel ground wire. (A) Short line. (B) Long line [14].

In accordance with the presented diagrams, the considered impedance is smaller in the case of the transmission line equipped with more conductive ACSR ground wire.

Now, let us assume that the impedances are $Z_A^+ = Z_A^- =$ j3.6 Ω and $Z_A^0 =$ j2.1 Ω, or adopted so that the ground fault current values are within the range of those found in most actual power systems. By using these and already given data, the results obtained on the basis of the sequential (tower-to-tower) ground fault analyses are displayed in Fig. 5.9.

To acquire a better insight into the considered effects, current I_e in the diagram is presented in relation to the total ground fault current for a fault assumed in substation A, $I_f(0)$. With the aim of determining the earth current relevant for the maximal step and touch voltages, obtained current

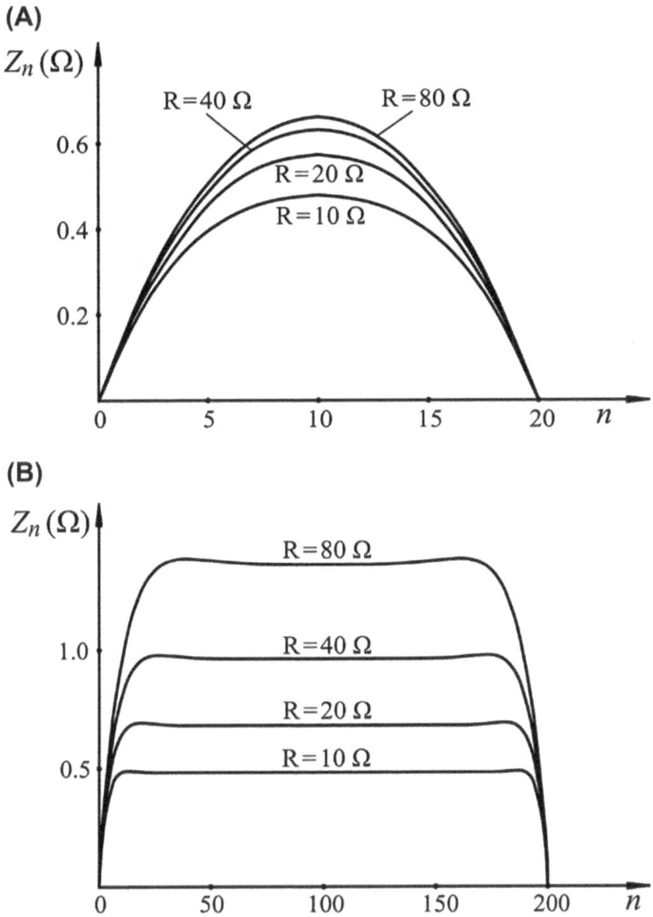

Figure 5.8 Impedance Z_n in the case of a line equipped with the ACSR ground wire. (A) Short line. (B) Long line [14].

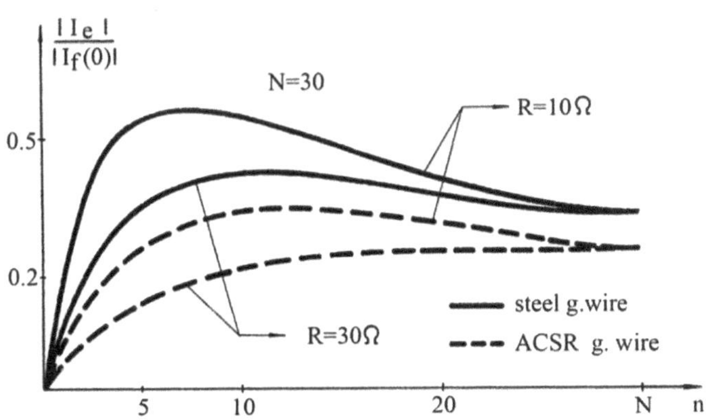

Figure 5.9 Current I_e as a function of fault location along the line [1].

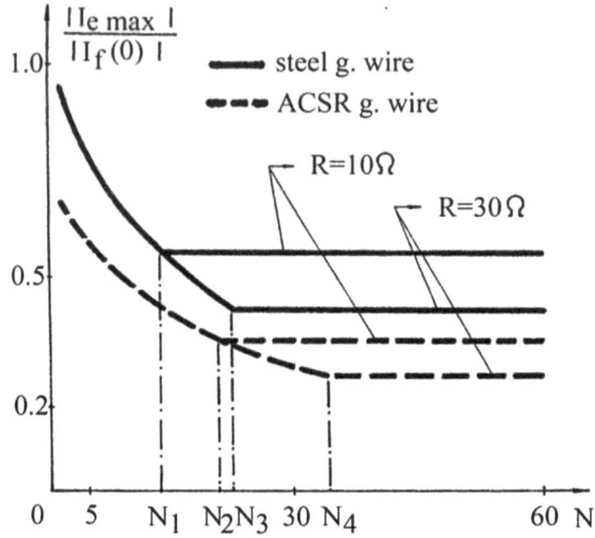

Figure 5.10 Current I_{emax} as a function of line length [1].

$I_e(c)$ should be compared with the largest earth current obtained for the fault in the considered generating substation itself, but in the higher voltage side.

On the basis of Fig. 5.9 it is possible to see that in the case of an ACSR ground wire and $R = 30\ \Omega$, current I_e does not reach its largest value for any fault along the line, but just when the fault occurs at the end of the line. Therefore it was necessary to investigate the position of the critical fault as a function of line length. The results of these calculations are displayed in Fig. 5.10.

The diagrams of Fig. 5.10 show that, for all considered cases, up to a certain limiting line length (N_1, N_2, N_3, and N_4) the largest value of the current occurs when the fault is at the end of the line (substation *B*). When the line length exceeds this length limit, current $I_e(c)$ remains practically constant with further increase in line length. At the same time, the critical fault position appears along the line at a certain distance from substation *A*, which also does not change with further increase in line length.

On the basis of the performed quantitative analysis, it is not difficult to see that among different outgoing lines the largest value of current $I_e(c)$ can be expected at those that are equipped with the less conductive ground wire(s) and at those that have lower tower footing resistances when all other relevant parameters are identical.

5.2.5 Validation of the Method

By applying the data used for obtaining the results shown in Fig. 5.9, the calculation was performed by utilizing a more rigorous, numerical method (the computer program based mainly on the procedure presented in Ref. [3]). The most important results from both calculations are presented in Table 5.1.

Comparison of the presented results shows that the error of the calculated value of current $I_e(c)$ in all the considered cases is negligibly small (less than 5%). These small errors may be explained as follows: total value of current I_f is determined by using symmetric components, i.e., by applying the approximation that the transmission line is absolutely symmetric. Since the configuration shown in Fig. 5.6 was intentionally chosen to emphasize the effects of this approximation, for most of the other possible line configurations, errors would be somewhat smaller compared to the ones shown in Table 5.1.

The presented analytical method gives a quick and, for practical purposes, sufficiently accurate evaluation of the significant parts of the ground fault current, for a fault at any of the towers of a transmission line of an arbitrary number of spans. The merit of the method in comparison with the existing numerical methods is the simple calculation procedure. The derived analytical expressions are obtained by applying the general equations of lumped parameter ladder circuits developed in Section 2.1 to the specific electrical circuit formed by a transmission line ground wire(s) under ground fault conditions.

The presented analytical method is convenient for ground fault analyses necessary for correctly designing grounding systems in cases where faults outside an HV genereting substation need to be considered. The method can also be used for selecting the ground wire capable of withstanding the

Table 5.1 Critical current passing through the station grounding system [1]

		Computer program			Method presented here		
Ground wire	**R(Ω)**	**$I_f(c)$ (kA)**	**$I_e(c)$ (kA)**	**$\frac{I_f(c)}{I_e(c)}$**	**$I_f(c)$ (kA)**	**$I_e(c)$ (kA)**	**$\frac{I_f(c)}{I_e(c)}$**
Steel	10	13.78	11.64	0.845	13.24	11.32	0.855
	30	11.05	9.11	0.825	10.46	8.69	0.831
ACSR	10	12.05	7.14	0.593	11.82	7.22	0.610
	30	10.24	5.60	0.547	9.92	5.54	0.558

anticipated fault current, as well as for evaluating safety conditions in the vicinity of certain transmission line towers. Thus the following can be concluded: by developing the presented methods, the number of practical problems that can be solved by relatively simple calculation procedures (without the use of computers) is enlarged.

5.3 GROUND FAULT IN A DOUBLE-CIRCUIT PARALLEL LINE

5.3.1 Introduction

The problem of determining the earth current I_e has been considered in the previous case of single-circuit lines. It is shown that in the case of single-circuit lines the largest (critical) value of this portion of the ground fault current can appear at the end of a relatively short line or for transmission lines of usual length at certain critical distances from the considered generating substation. Here the previously presented method has been extended to include the case when at least one of the outgoing lines is a double-circuit parallel line. The reason for focusing our attention on this type of transmission line is that higher ground fault currents can be expected in such lines compared to those of single-circuit lines, when all other conditions are the same or very similar. The place of fault is fed by the fault current from the considered generating substation via two lines (faulted and healthy line). Because impedance of the fault loop is in this way reduced in comparison with the single-circuit lines under similar conditions, higher total ground fault currents are to be expected in the case of double-circuit parallel lines.

Double-circuit three-phase lines are becoming increasingly frequent in HV transmission networks and it is a realistic assumption than in the future their number will increase further. Since they actually represent two parallel lines that share the same route (towers), the specific property of these lines is that they are electromagnetically coupled. The influence of this coupling can be neglected in the case of symmetric operation with positive or negative sequence, but it is evident in the zero sequence operation [16]. In the case of inductively coupled lines the value of the zero impedance of each line seen separately is increased, and in the equivalent circuit of these lines mutual zero sequence impedances appear as new ones in comparison with the former case. In the method presented here the increased zero sequence impedance of the parallel lines is involved by using the analytical expression given in Ref. [17], whereas the influence of the mutual impedance is appreciated by using the decoupling technique. Since the

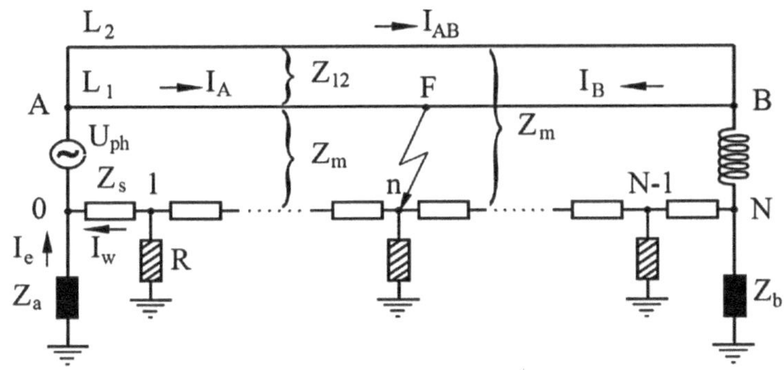

Figure 5.11 Parts of the ground fault current [13].

spatial disposition of the conductors of these lines is symmetrical (Fig. 5.16), the relevant parameters of both lines are practically identical. Such configuration of this type of transmission line certainly represents a convenience that enables derivation of the relatively simple analytical expressions, regardless of the doubled number of phase and neutral conductors in comparison with the formerly considered case of single circuit lines.

5.3.2 Ground Fault at an Arbitrary Place

Let us consider a simple power system consisting of a double circuit parallel line through which an HV source substation, *A*, supplies one distribution, substation *B*. Thus all six phase conductors stem from the same substation, share the same towers, and terminate at the same substation as shown in Fig. 5.11.

The used notation has the following meaning:

U_{ph}—phase voltage at the source station,

L_1, L_2—faulted and healthy line, respectively,

$I_A(I_B)$—fault current component through the faulted line to the left (right) of the fault location, *F*,

I_{AB}—fault current component through the healthy line,

I_e—earth current through the grounding system of source station *A*,

I_W—fault current through the ground wire(s) in the first span, counted from station *A*.

$Z_a(Z_b)$—impedance of the grounding system of station *A* (*B*), which does not include grounding effects of the ground wire(s) of the line under consideration,

Z_s—self-impedance of the ground wire(s) for the length of one average span,

Z_{12}—mutual impedance between the phase conductor carrying current I_{AB} and the phase conductor carrying currents I_A and I_B,

Z_m—mutual impedance between the ground wire(s) and the faulty phase conductor, as well as between the ground wire(s) and the phase conductor carrying the fault current, I_{AB} (L_2),

R—average tower footing resistance,

n—number of spans to the fault location, counted from substation A, and

N—total number of spans.

The circuit presented in Fig. 5.11 shows parts of the fault current when a single line-to-ground fault occurs at an arbitrary (nth) tower. Impedances Z_s, Z_{12}, and Z_m are determined by the well-known calculation procedure described in, e.g., Refs. [13,15].

As can be seen from Fig. 5.11, double-circuit parallel lines have two conductors carrying the ground fault current and according to this two mutual impedances between the ground wire(s) and these phase conductors. Since the position of these conductors relative to the ground wire(s) is symmetrical, as can be seen in Fig. 5.16 (phase conductors C_1 and C_2), these impedances have the same value. This fact makes it possible to use the same notation for each one of them (Z_m). Since current I_e is to be determined under the worst fault conditions we will assume that the faulted conductor is the most distant from the ground wire(s).

The specificity of double-circuit lines is also mutual impedance Z_{12}. In nontransposed lines, usually relatively short ones, their value is constant, since positions of the phase conductors (C_1 and C_2) along such lines do not change. However, when the lines are transposed, this impedance changes their value at each transposition point since the space position of all phase conductors changes from A_1 A_2 to B_1B_2, from B_1B_2 to C_1C_2, and from C_1C_2 to A_1A_2 (Fig. 5.16). However, a change in the relative position of the phase conductors of a transposed line under practical conditions does not influence significantly the value of impedance Z_m, since its dependence on this distance is logarithmic. Therefore for practical purposes the average value of the mentioned distances can be adopted for the necessary calculations. However, since current I_e is to be determined under the worst fault conditions we have to assume that the faulted conductor is the most distant from the ground wire(s). In cases of transposed lines, we also ensure that the finally obtained results are slightly on the safe side.

The many so far performed quantitative analyses of various line configurations show that the inductive influence on other conductors of both lines (L_1 and L_2) can be neglected even when it is at its maximum (fault at the end of the line). This can be explained by the fact that the currents induced in these conductors at the ends of the line circulate through power transformer windings having high inductive impedance. Thus these currents can be neglected. This approximation also simplifies the problem and, at the same time, guarantees slightly conservative results for the total ground fault current I_f and its parts.

As in the previous case it is necessary to determine current I_e at the critical distance, when it reaches its maximum. For this purpose the power system that we have adopted (shown in Fig. 5.1) can be initiated by using:

- the method of symmetric components,
- the driving point technique,
- the decoupled technique, and
- reduction of the uniform lumped parameter ladder circuit, Chapter 2, presented by the equivalent circuit in Fig. 5.12.

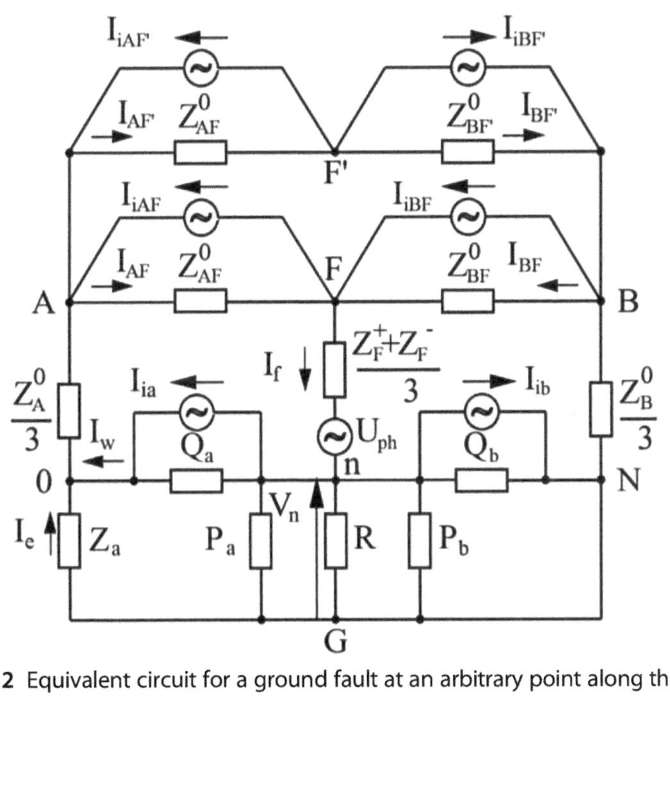

Figure 5.12 Equivalent circuit for a ground fault at an arbitrary point along the line [13].

The used notation has the following meaning:
U_{ph}—driving voltage source at the place of the fault, whose voltage is equal to the phase voltage at the source substation,
V_n—potential of the faulted tower,
I_f—total ground fault current,
$I_{AF'}$, $I_{iBF'}$—current sources substituting the inductive influence of the faulted line on the healthy line,
I_{AF}, I_{iBF}—current sources substituting the inductive influence of the healthy line on the faulted line,
I_{ia}, I_{ib}—current sources substituting the inductive influence of the phase conductors on the ground wire(s),
Z_F^+, Z_F^- —positive and negative sequence impedances of the system at the fault place,
$Z_A^0\left(Z_B^0\right)$—zero sequence impedance of the system at substation A (B),
Z_{AB}^0—zero sequence impedance of the healthy line,
$Z_{AF}^0\left(Z_{BF}^0\right)$—zero sequence impedance of the line from the fault location to substation A (B), and
G—remote ground.

Current sources $I_{iAF'}$, I_{iAF}, $I_{iBF'}$, and I_{iBF} appear as the consequence of the mutually inductive coupling of the phase conductors and are determined by the following relations:

$$I_{iAF'} = \mu I_A = \mu(I_{AF} - I_{iAF}), \tag{5.33}$$

$$I_{iAF} = \mu I_{AB} = \mu(I_{AF'} - I_{iAF'}), \tag{5.34}$$

$$I_{iBF'} = \mu I_B = \mu(I_{BF} + I_{iBF}), \text{ and} \tag{5.35}$$

$$I_{iBF} = \mu I_{AB} = \mu(I_{BF'} + I_{iBF'}). \tag{5.36}$$

In these expressions, I_{AF} and I_{BF} represent the currents appearing in the faulted phase conductor due to the galvanic connection between A and F and B and F, respectively. Currents $I_{AF'}$ and $I_{BF'}$ appear in the corresponding phase conductor of the healthy line because of the galvanic connection between points A and F' and B and F', respectively. Point F' is an imagined point on the healthy line, determined by the position of the faulted tower. Currents $I_{iAF'}$ and $I_{iBF'}$ are induced in the phase conductor of the healthy line by currents I_A and I_B, respectively. On the other hand, I_{iAF} and I_{iBF} are the currents induced in the faulty conductor by current I_{AB} in sections AF and BF, respectively. The symbol μ denotes the coefficient of inductive coupling between the two ground fault current circuits with common earth return path, determined by:

$$\mu = \frac{Z'_{12}}{Z'_{ph}}, \tag{5.37}$$

where Z'_{ph} is the self-impedance of one of the phase conductors carrying the ground fault current per unit length and Z'_{12} is the mutual impedance between the phase conductors carrying the ground fault current per unit length.

Currents induced in the ground wire(s) are represented by the current sources whose currents are:

$$I_{ia} = \frac{Z'_m}{Z'_{gw}}(I_A + I_{AB}), \tag{5.38}$$

$$I_{ib} = \frac{Z'_m}{Z'_m}(I_B - I_{AB}), \tag{5.39}$$

where Z'_{gw} is the self-impedance of the ground wire(s) per unit length and Z'_m is the mutual impedance between the ground wire(s) and one of the phase conductors carrying the ground fault current per unit length.

As in the case of single-circuit lines, equivalent impedances Q_a and P_a are determined by expressions (5.3) and (5.4). Impedances Q_b and P_b can also be determined by expressions (5.3) and (5.4); it is only necessary to introduce the corresponding number of spans, i.e., to replace n by $m = N-n$.

The equivalent circuit and the necessary calculations can be further simplified if the section from the fault point F to substation B can be considered infinitely long from the standpoint of the grounding effects, i.e., if the criterion defined by Eq. (5.19) is satisfied.

Impedances Z_F^+ and Z_F^- are determined by using the equivalent circuit shown in Fig. 5.13, while the analytical expressions for the positive, negative, and zero sequence line impedances are given in Refs. [17,18].

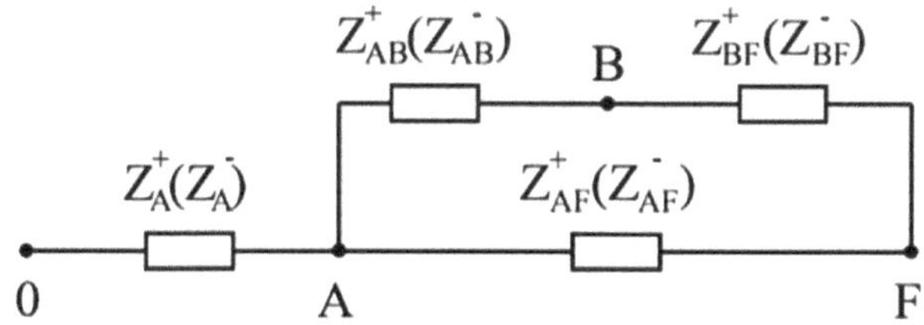

Figure 5.13 The equivalent circuit for determination of Z_F^+ and Z_F^- [13].

The used notation has the following meaning:

Z_A^+, Z_A^-—positive and negative sequence impedances of the electric power system at station A,

Z_{AB}^+, Z_{AB}^-—positive and negative sequence impedances of one line from A to B,

$Z_{AF}^+(Z_{BF}^+)$—positive sequence impedance of one line from F to A (B), and

$Z_{AF}^-(Z_{BF}^-)$—negative sequence impedance of one line from F to A (B).

For the purpose of calculation of current I_e in the case of a fault at the end of the line or in substation B ($n = N$), the equivalent circuit shown in Fig. 5.12 becomes significantly simpler. The following impedances are then omitted from the equivalent circuit in Fig. 5.12: Q_b, P_b, and Z_{BF}^0, as well as resistance R. Also impedance Z_{AF}^0 becomes equal to impedance Z_{AB}^0. Finally, all of the four current sources representing the currents induced in the phase conductors can be replaced by two mutually equal current sources.

On the basis of the equivalent circuit of Fig. 5.12 and the known analytical expressions for determining its impedances (e.g., Ref. [18]), the corresponding system of equations can be completed. Then, by successively increasing n and by performing sequential (tower-to-tower) calculations, it is possible to determine the largest possible values of current I_e, current I_w, and potential of any tower, V_n, of a designed double-circuit parallel line.

The described procedure can be used for a double-circuit parallel line of an arbitrarily complex power system, which implies multiple lines, multiple sources, multiple substations, different voltage levels, substations with directly grounded neutral point(s), etc. Prior to this it is only necessary to calculate the positive, negative, and zero sequence impedances of the system at the terminal points of the considered line. Also to determine I_e, the equivalent circuit should be formed in such a way to retain one branch that directly connects points A and G (Fig. 5.11). The equivalent impedance in this branch should be determined so that the fault current is divided into two constitutive components. One of them circulates through the neutral(s) in substation A, while the other flows through the remaining grounded neutral point(s) in the network, with the exception of the part of the network supplied by the considered line.

5.3.3 Critical Fault Position

With the aim of avoiding numerous successive (tower-to-tower) ground fault analyses the equivalent circuit shown in Fig. 5.12 can be modified to become significantly simpler. First, in many practical cases the condition

$Z_A^0 \ll Z_B^0$ is satisfied and the branch with impedance $Z_B^0/3$ can be omitted. As a consequence, according to Fig. 5.12, currents I_B and I_{AB} become the same current. Also current sources I_{iBF} and $I_{iBF'}$, according to Eqs. (5.34) and (5.35), become mutually equal. Finally, influences of each of these sources on the value of the current outside the phase conductors cancel each other (Fig. 5.12), so they can also be omitted. Thus the equivalent circuit of Fig. 5.12 modified in accordance with the previous approximations obtains the appearance as shown in Fig. 5.14.

As in the case of single-circuit lines, the grounding impedance at an arbitrary fault location, Z_n, in the equivalent circuit in Fig. 5.13 is determined by expression (5.14). Also the extensive (tower-to-tower) quantitative analysis of the influence of inductive coupling of the phase conductors of the two lines shows that it has a very small influence on the critical fault position. In only a few of the considered cases, by taking into account this influence, does the critical fault position become nearer to the generating substation by one span. This practically means that in determining the critical fault position, c, current sources I_{iAF} and $I_{iAF'}$ can be omitted without significantly affecting calculation accuracy. Thus the

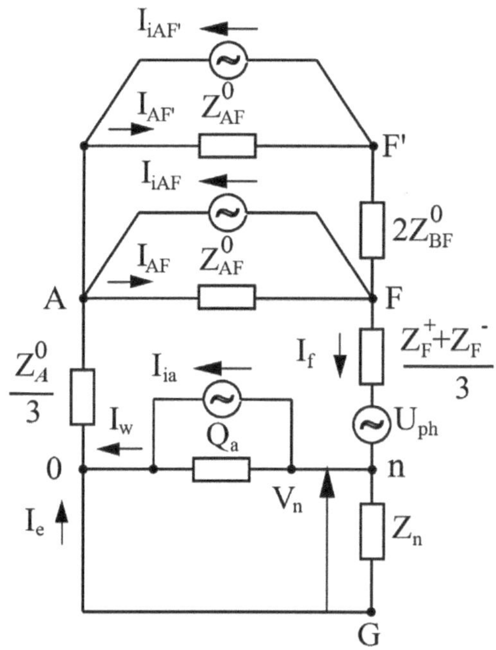

Figure 5.14 The simplified equivalent circuit [13].

whole problem is reduced to the previously considered case of single-circuit lines. Analogous to the previously presented procedure, but now taking into account impedance Z_a, we obtain the following approximation:

$$c + \frac{2Z_A + r(Z_\infty + 2Z_a)}{2\Delta Z_c} - \frac{R}{Z_\infty} \approx \frac{(k^{2c+1} - 1)R}{k[k^c(Z_\infty + 2Z_a) + Z_\infty]}, \tag{5.40}$$

where:

$$Z_A = \frac{Z_A^+ + Z_A^- + Z_A^0}{3}, \tag{5.41}$$

$$r = 1 - \frac{Z'_m}{Z'_{gw}}. \tag{5.42}$$

Impedance ΔZ_c represents an increment of the impedance of the fault loop at the critical fault position if the line length is increased by one span. On the basis of this definition, according to the equivalent circuits shown in Figs. 5.13 and 5.14 (without current sources I_{iAF} and $I_{iAF'}$), we obtain:

$$\Delta Z_c = \frac{(c+1)[2N - (c+1)]}{2N}\Delta Z - \frac{c(2N - c)}{2N}\Delta Z. \tag{5.43}$$

Then, after certain algebraic manipulation, Eq. (5.43) becomes:

$$\Delta Z_c = \frac{2N - 2c - 1}{2N}\Delta Z, \tag{5.44}$$

where:

$$\Delta Z = \frac{2Z_S^+ + Z_S^0}{3}. \tag{5.45}$$

Impedances Z_S^+ and Z_S^0 are the positive and zero sequence impedances per one span of one of the parallel lines (Fig. 5.16).

Finally, at the design stage, it is usually assumed that impedance Z_a is negligibly small, and then instead of Eq. (5.40) we can use:

$$c + \frac{2Z_A + rZ_\infty}{2\Delta Z_c} \approx \frac{k^c}{k-1} + 1. \tag{5.46}$$

By using approximation (5.46) the estimated critical fault position becomes nearer to substation A and, as a consequence, the determined maximum of current $I_e(c)$ is somewhat larger. Also it is necessary to stress the following: since impedance Z_a is also neglected in the equivalent circuits (Figs. 5.12 and 5.14), the estimated critical value of current I_e for cases where impedance Z_a has a high value (e.g., about 1 Ω) will be somewhat larger than the actual one. However, both of these approximations should be considered as rather desirable because they lead to the final results that are slightly shifted to the safe side.

Approximations (5.40) and (5.46) represent a transcendental equation whose solution may be obtained only graphically. The diagrams shown in Fig. 5.15 represent an illustration of the procedure necessary for obtaining the solution to Eq. (5.46) for one of the cases studied in the quantitative analysis presented in this section.

Fig. 5.15 graphically shows functions representing the right-hand side (F_r) and the left-hand side of Eq. (5.40) for different line lengths: F_1 for $N = 20$, F_2 for $N = 40$, and F_3 for $N = 200$. The points c_1 and c_2 represent the positions of critical faults for lines of 40 and 200 spans, respectively. Certainly, since c can only be an integer, the final solution must be sought among the nearest integers, making use of the equivalent circuit shown in Fig. 5.14.

If the curve representing the left-hand side of Eq. (5.46) or (5.40) does not intersect with curve F_r representing the right-hand site of this equation and instead becomes divergent with an increase in n (e.g., curve F_1), function $I_e(n)$ does not reach its maximum, and the largest/critical value of I_e is obtained for a fault at the end of the line. Thus the presented graphical solution helps us to reduce additionally the calculation procedure described in the case of single-circuit lines. Namely, the solution to Eq. (5.40) presented in Fig. 5.15 gives us simultaneously a quick answer to the

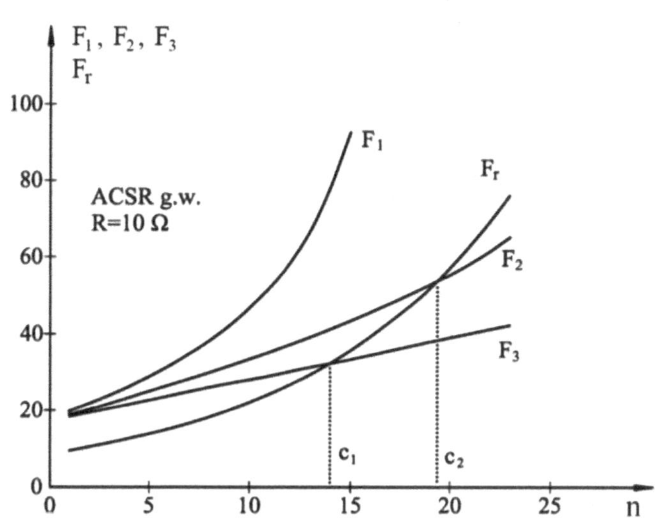

Figure 5.15 Determination of the critical fault position [13].

question: does current I_e reach its largest value for a fault at a certain critical distance, c, or at the end of the line? Thus the presented procedure eliminates the need for the procedure to determine the limiting line length described in the previously considered case of single-circuit lines.

5.3.4 Quantitative Analysis

Let us assume that a double-circuit line connecting stations A and B is a nontransposed 110 kV line with six aluminum/steel 240/40 mm^2 phase conductors, configured as shown in Fig. 5.16.

The ground wires are in one case steel 50 mm^2 and in the other case ACSR 95/55 mm^2. Line impedances per one span are determined on the basis of the following assumptions: average length of a span is 250 m and specific soil resistance along the line is 50 Ωm. The resistances per unit length of phase conductors, steel, and ACSR ground wire are 0.12, 3.5, and 0.299 Ω/km, respectively.

Source impedances $Z_A^+ = Z_A^- = \mathrm{j}3.6\ \Omega$ and $Z_A^0 = \mathrm{j}2.1\ \Omega$ are chosen so that the values of the ground fault currents are within the range of those found in actual power systems. Substation B is with the insulated neutral point(s) and impedances Z_a and Z_b are so small that they can be considered equal to zero.

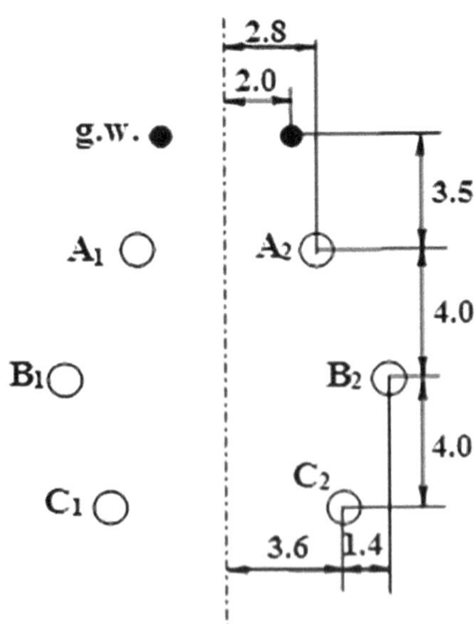

Figure 5.16 The line configuration [13].

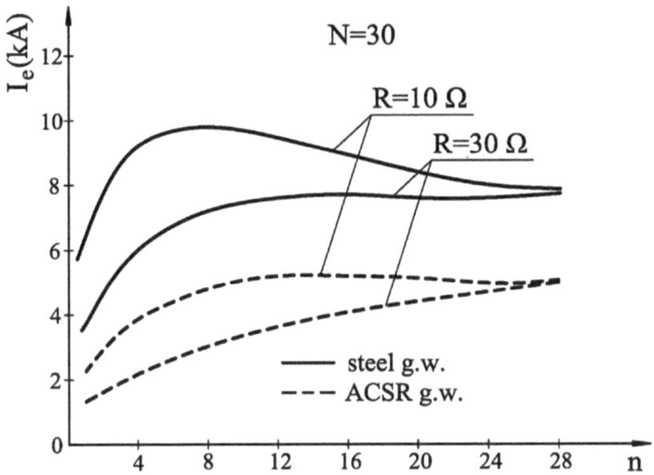

Figure 5.17 Current I_e as a function of the fault place along a short line [13].

On the basis of the given data, by using the described procedure we have obtained the results presented in Figs. 5.17 and 5.18.

As can be seen from the diagrams in Fig. 5.17, change of current I_e depending on the fault location is considered in a relatively short line, $N = 30$. For both types of adopted ground wires, if $R = 30\ \Omega$, current I_e does not reach its largest value for any place along the line, but at the end of the line, in supply substation *B*.

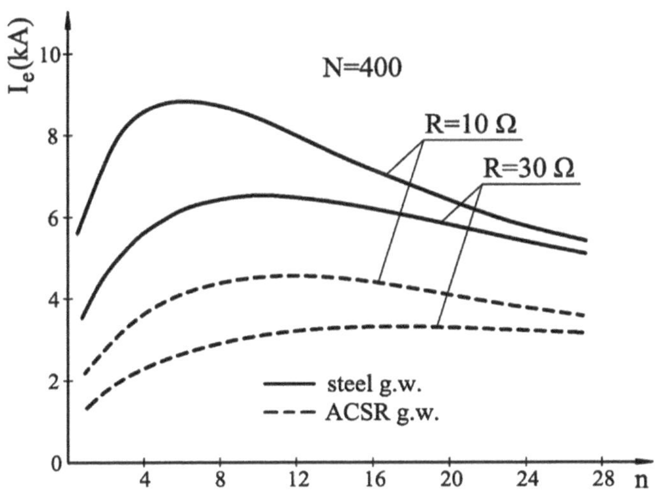

Figure 5.18 Current I_e as a function of the fault place along a long line [13].

As is shown in Fig. 5.18, in the case of a relatively long line, $N = 400$, current I_e in all four considered cases reaches its maximum for a fault at certain place along the line, but its intensity decreases in comparison with the formerly considered case ($N = 30$), since the contribution of the fault current via the healthy line (I_{AB}) is decreased with increased length of the line. Also in the case of a long line the position of critical fault c shifts toward the source substation (also seen in Fig. 5.15), and the limiting point of this shift is reached if the line becomes so long that the contribution of I_{AB} to current $I_e(c)$ can be neglected.

The influence of inductive coupling of the phase conductors to the decrease in current $I_e(c)$, expressed in relative units, amounts to 8% in most cases of steel ground wire(s) or 10% in cases of ACSR ground wire(s). In most practical analyses this could be neglected, which would greatly simplify the circuits of Figs. 5.12 and 5.14. This influence is more pronounced at the end of the line and increases with an increase in line length, asymptotically approaching a certain limit value. Thus for making a quick and easy preliminary estimation, in most practical cases this influence can be also disregarded and the circuit in Fig. 5.14 can be additionally simplified.

On the basis of the performed quantitative analysis it is not difficult to see that the critical fault position, c, assuming all other conditions are identical or similar, can be expected in the case of a double-circuit parallel line rather than in the case of a single-circuit line, especially in the case of relatively short lines. Among different outgoing double-circuit parallel lines the largest value of current $I_e(c)$ can be expected at those that have a less conductive ground wire(s), as well as at those that have the lowest tower footing resistances, if all other relevant parameters are identical or similar.

The presented practical method gives a quick and, for practical purposes, sufficiently accurate evaluation of the main parts of the ground fault current for ground faults at any of the towers of a double three-phase circuit line with an arbitrary number of spans. The simplicity of the method originates from the application of relatively simple and exact equations for uniform lumped parameter ladder circuits of arbitrary size (for any number of discrete parameters) and for any terminal conditions (Section 2.1). The method can be applied to solve several practical problems: evaluation of the largest value of current I_e, selection of a ground wire capable of withstanding fault currents, and prediction of the step and touch voltages in the vicinity of any of the double-circuit parallel line towers.

The presented method is applicable in the analysis and design of grounding systems of HV generating substations having at least one outgoing double-circuit parallel line and in the design of double-circuit parallel lines. The approximations and assumptions made are those that we are otherwise forced to adopt at the design stage because of the uncertainty of the basic data of soil resistivity at each designed transmission tower.

The presented analytical procedure represents a practical (directly applicable) method for determining critical earth current that will pass through the grounding system of a designed generating substation. The grounding systems of these substations involve, as is well known, the substation grounding grid and many tower footing electrodes belonging to the connected transmission and distribution lines. Since these electrodes are disposed at great mutual distances between them, there is no mutual interference in performing their grounding function. Thus in many practical cases the grounding impedances of this type of HV substation are determined by using relatively simple analytical expressions usually given in the corresponding technical standards, e.g., Ref. [19].

The presented method certainly enlarges the number of practical problems that can be solved by using relatively simple calculation procedures.

REFERENCES

[1] Lj.M. Popović, Practical method for evaluating ground fault current distribution in station, towers and ground wire, IEEE Transactions on Power Delivery 13 (1) (1998) 123–128.

[2] R. Varma, D. Mukhedkar, Ground fault current distribution in substation, towers and ground wires, IEEE Transactions on Power Apparatus and Systems PAS 98 (3) (1979) 724–730.

[3] A. Meliopoulos, R. Webb, E. Joy, S. Patel, Computation of maximum earth current in substation switchyards, IEEE Transactions on Power Apparatus and Systems PAS 102 (9) (1983) 3131–3139.

[4] D. Garrett, J. Myers, S. Patel, Determination of maximum substation grounding system fault current using graphical analysis, IEEE Transactions on Power Delivery PWRD 2 (3) (1987) 725–732.

[5] F. Dawalibi, G. Niels, Measurements and computations of fault current distribution on overhead transmission lines, IEEE Transactions on Power Apparatus and Systems PAS 103 (3) (1984) 533–560.

[6] H. Goci, S. Sebo, Distribution of ground fault current along transmission lines - an improved algorithm, IEEE Transactions on Power Apparatus and Systems PAS 104 (3) (1985) 663–669.

[7] G. Niles, et al., Background and methodology for analyzing step and touch potentials near transmission structures, IEEE Transactions on Power Systems PWRD 1 (2) (1986) 150–157.

[8] S. Lambert, Minimum shield wire size - fault current considerations, IEEE Transactions on Power Apparatus and Systems PAS 102 (3) (1983) 572–578.
[9] S.T. Sobral, V.S. Costa, M.S. Campas, D. Mukhedkar, Dimensioning of nearby substations interconnected grounding system, IEEE Transactions on Power Delivery 3 (4) (1988) 1605–1614.
[10] S. Sebo, Zero sequence current distribution along transmission lines, IEEE Transactions on Power Apparatus and Systems PAS 88 (6) (1969) 910–919.
[11] B. Thapar, S. Madan, Current for design of grounding systems, IEEE Transactions on Power Apparatus and Systems PAS 103 (9) (1984) 2633–2636.
[12] Lj.M. Popović, Dissipating of the partial ground fault current across the shield wires of transmission lines, Electronic Power Systems Research 11 (1) (1986) 25–37.
[13] Lj.M. Popović, A practical method for evaluation of ground fault current distribution on double circuit parallel lines, IEEE Transactions on Power Delivery 15 (1) (2000) 108–113.
[14] Lj.M. Popović, Analytical expression for voltage drops on return paths of ground fault current when fault occur in transmission line, in: IEEE PES Summer Meeting, 2001 (Conference Proceedings).
[15] Short-Circuits Currents in Three-Phase A.C. System—Part 3: Currents During Two Separate Simultaneous Line-to-earth Short Circuit and Partial Short-circuit Flowing Trough Earth, IEC Int. Std. 60909-3, Ed. 3.0, 2009-03.
[16] Lj.M. Popović, General equations of the line represented by discrete parameters, part I – steady state, IEEE Transactions on Power Delivery 6 (1) (1991) 295–301.
[17] Technical Report IEC 909-2, Electrical Equipment" – Data for Short-Circuit Current Calculations in Accordance with IEC 909, August 1992.
[18] W.H. Kersting, Distribution System Modeling and Analysis, CRC Press LLC, USA, 2002.
[19] IEEE Guide for Safety in Substations Grounding, Std. 80 –2000.

CHAPTER 6

Grounding Systems Consisting of Long External Electrodes

6.1 INTRODUCTION

This chapter is devoted to the most complex grounding systems that can be encountered in practice. These are the grounding systems of high-voltage (HV) distribution substations located in urban or suburban areas. The contemporary concept in solving this type of grounding problem is based on the grounding effects of all outgoing metal installations no matter what their basic function is. They are metal sheaths of connected cable lines having uncovered or covered outer sheaths, ground wires of connected overhead lines, and all other types of metal installations in direct and continuous, or indirect and sporadic, contact with the earth. Their common characteristic is that they emerge from the yard of an HV substation and pass through a wide urban area around it. According to this we will use the term external grounding electrodes for all these types of metal installation.

It is well known that medium-voltage (MV) paper-insulated cable lines act as long grounding conductors, i.e., they dissipate fault current directly into the surrounding soil. As a consequence, these cable lines spontaneously form a very large and complex grounding system around HV/MV distribution substations. The area encompassed by this type of grounding system is much larger than the one occupied by an HV/MV substation alone. Because of extremly large involved area these grounding systems manifest excellent characteristics before all a very low value of the grounding impedance. The problem with this type of grounding system appears only at the design stage of an HV/MV substation, when safety conditions inside and in the vicinity of this substation should be estimated and confirmed by calculation. The reason for this is that these grounding systems have a very complex structure and configuration, which are mainly determined by local urban surroundings and because of that are different for each concrete case.

Grounding systems of HV/MV distribution substations are not strictly limited to a priori defined areas. They are formed freely and spontaneously

Practical Methods for Analysis and Design of HV Installation Grounding Systems
ISBN 978-0-12-814460-2
https://doi.org/10.1016/B978-0-12-814460-2.00006-4

so that they encompass many various underground metal installations and constructions, including those that only in a trivial way, because of the great distance from the HV/MV substation (the source of the razed ground potential), perform their grounding function, i.e., dissipate a ground fault current into the surrounding earth. According to the considerations given in Section 2.2 (Fig. 2.8), spatial limits of these grounding systems exist only because the cable sheath transfer impedances to the certain distances seen from the HV substation become practically infinitely large. Such large impedances block further axial flows of the fault current through the cable sheaths (Fig. 2.18). In other words, the mentioned limits exist because the outgoing MV cable lines can perform their grounding function only along a certain limited length, the so-called active cable line length. Beyond this length the transfer (axial) impedance of the cable line sheath becomes practically infinitely large and the cable line cannot perform its grounding function, or dissipate ground fault current into the surrounding earth. Such a grounding system, certainly, involves the largest possible area, and for the certain constructive characteristics of the implemented MV cables it is dependent only on resistivity of the soil of the involved urban area. Thus it can be said that urban areas encompassed by a grounding system formed in this way are practically identical to the potential field produced during a ground fault on the earth surface by the grounding system itself. Certainly, such grounding systems manifest excellent characteristics and the problem arises only when this fact should be confirmed by the calculations required at the design stage of an HV substation.

The complexity of the problem caused by many relevant factors as well as by uncertainty and deficiency of necessary data affected the development of necessary calculation methods. Thus through their development, two different tendencies can be recognized. On the one hand, by applying reasonable assumptions and idealizations of the real physical model, efforts have been made to develop directly applicable and relatively simple calculation procedures, e.g., Refs. [1—8]. On the other hand, research efforts have been made to ensure that the developed mathematical model takes into account, as accurately as possible, as many relevant factors and parameters, including those of the lower order, e.g., Refs. [9—11].

The only analytical method for the analysis and design of this type of grounding system known so far is presented here on the basis of Refs. [1—5]. By using this method designers of an HV distribution substation have at their disposal an appropriate formalism for describing concisely the grounding contribution of all components of a grounding system. This formalism has the form of analytical expressions enabling evaluation of the grounding contribution of each grounding system component seen separately.

A computational method can also be used for solving this type of grounding problem, e.g., Refs. [9,11]. According to these methods the outgoing cable lines are fictitiously divided into many segments to model their grounding effects and voltage drops along the individual cable metal sheath. Each segment is characterized by its self-inductance and mutual inductances with respect to other segments of the same cable line and also of other cable lines, as well as by its resistance with respect to the ground and mutual resistance regarding the remaining cable's segments and the elements of the HV/MV substation grounding electrode (grid). The favorable property of this method is that it offers the possibility of considering the influence of the factors of lower order. However, in many concrete cases, the high accuracy of these methods cannot be utilized because of the practical difficulties appearing at the design stage of collecting all input data necessary for obtaining final results of high accuracy.

Obviously, it is necessary to find a compromise between the two opposite requirements: one directed toward greater accuracy and other directed toward greater simplicity of the necessary calculation procedure. Both of these requirements are satisfied to a great extent by the method finally developed in Refs. [1,2]. It is developed under the following assumptions:

- The grounding contribution of the substation grounding electrode is calculated as if this is the only existing element of the whole grounding system.
- The grounding contributions of individual outgoing cable lines are calculated in one case depending on their space dispositions in relation to the substation grounding electrode and in the other depending on their space disposition in relation to the other cable lines.

According to such an approach, all outgoing cable lines with uncovered metal sheaths are sorted into two characteristic groups:

- Cable lines laid over the entire active length in one of the few (usually four) primary directions, and
- Cable lines that at some distance less than their active length divert from the primary direction into an essentially different direction, commonly called the secondary direction.

The grounding contribution of a cable line belonging to the first group is calculated by using an analytical expression that takes into consideration mutual interference between this cable line and the substation grounding electrode in dissipating the fault current into the surrounding earth. The grounding contribution of a cable line belonging to the second group is calculated by using an analytical expression that takes into account the mutual interference between this cable line and all cable lines belonging to the same primary direction. Mutual interference in dissipating a fault current into the earth, or the so-called proximity effect [3], between mutually distant elements, as if they were the cable lines belonging to the different primary directions, are disregarded.

When applying any of the mentioned methods, problems appear when data necessary for calculations are to be prepared/collected. One of main difficulties is because of the fact that the basic data, the value of the equivalent soil resistivity of the surrounding urban area covered by the HV/MV substation grounding system cannot be determined exactly. Thus substation designers are forced to adopt an approximately estimated value and use it for the necessary calculations. Therefore there is no practical purpose to insisting on a high accuracy of the calculation method, or its ability to take into consideration many factors of lower order. One can achieve only the approximate accuracy based on the so-called reasonable assumptions and approximations. Thus a developed method can be considered as sufficiently accurate if it enables correct appreciation of the influence of the factors that are of prime importance. When applying unavoidable approximations and idealizations of the real physical model, it is sufficient to know that their influences on the final calculation result are small and if it is possible slightly on the conservative side.

The analytical expressions presented here, although relatively simple, allow evaluation of the grounding contributions of the cable lines and grounding contributions of all other underground metal installations typical for urban conditions (different type of metal pipelines, steel reinforcement in the building foundations, etc.) connected to the grounding system through the low-voltage (LV) neutrals at the nearest customer (MV/LV) substations. Also the presented method is simple for application and suitable for obtaining a necessary insight into the influence and interplay of relevant factors and parameters. Due to this, the presented method is, above all, convenient and useful for those dealing with practical problems.

Having in mind the number and complexity of metal installations of urban areas covered by this type of grounding system and objective deficiency of many relevant data, it is no wonder that the problem of actual ground fault current distribution in the case of these grounding systems was for a long time treated as unsolvable [12–19]. Practical problems are still being solved without taking into consideration the influence of metal installations surrounding feeding lines passing through urban and/or suburban areas [20]. The only method enabling determination of the actual reduction factor of feeding lines passing through urban and/or suburban areas is presented in Section 3.2 on the basis of Refs. [12,13].

From the standpoint of ground fault current distribution it is also important to say that the essential difference exists between substations located in an urban area and those located in a suburban area. The former are exclusively supplied by cable lines, while the latter are frequently supplied by overhead lines. It is also important to say that the former are usually built as gas-insulated substations, while the latter are built as outdoor (conventional) substations. Also the substation grounding electrode in the former case is steel armature of the substation building foundation, while in the latter case it is constructed as a ground grid.

6.2 GROUNDING SYSTEM OF HV SUBSTATIONS LOCATED IN URBAN AREAS

6.2.1 Description of the Spontaneously Formed Grounding Systems

There are several reasons why grounding problems of HV substations located in urban or suburban areas, both conventional and gas-insulated, merit special consideration. First of all, the neutral points of the HV networks that supply these stations are, as a rule, solidly earthed, and so the ground fault currents reach very high values that can, in some cases, exceed 20 kA. Furthermore, for more efficient electric power distribution there is a tendency to place these stations at load centers, i.e., deep within urban zones, where there is of course very little free space for these substations and where numerous objects are accessible to the general public.

The fact that an HV substation is located in urban area means that the spatial disposition of the grounding grid and the large number of the outgoing cable lines is determined primarily by the disposition of the surrounding traffic lines (streets). Therefore the substation designer has fewer possibilities to influence the disposition of the grounding system elements. At first sight, this seems to imply that the configuration of the grounding system differs from case to case and that it is consequently impossible to generalize its configuration. However, even though the grounding system configurations are different in each concrete case, all of them can be

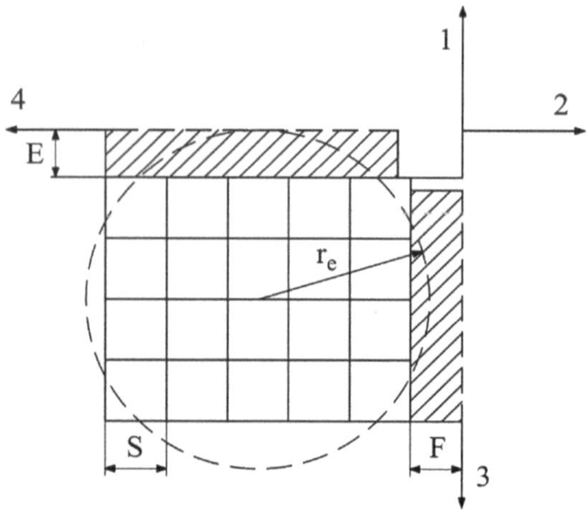

Figure 6.1 The ground grid and primary directions for installing cable lines [1].

decomposed into only three characteristic components, as presented in Figs. 6.1 and 6.2. These components are the following:

- Substation grounding electrode (grid),
- Cable lines leaving the substation yard and along their whole active length belonging to one of the few essentially different directions, called primary directions, and
- Cable lines leaving the substation area, and at a distance less than their active lengths, diverting from the primary direction into one essentially different direction, commonly called the secondary direction.

In a fully loaded substation, the number of outgoing MV cable lines varies from 30 to 40, but the number of directions in which they can be laid in the vicinity of the substation is most frequently limited to four, when the station is located at a crossroads of two urban traffic lines (streets) (Fig. 6.1).

The notation has the following meaning:

1, 2, 3, and 4—primary directions in which the cables can be laid,

E and *F*—distances of the cable line sections from the nearest parallel grid edge,

S—grid spacing, and

r_e—radius of the equivalent circular plate of an area equal to the area of the substation ground grid.

The broken straight line represents the section of directions 4 and 3 in which it is convenient to treat the cable lines as the grid elements, and, according to this to treat the shaded area as the increase in the area covered by the ground grid.

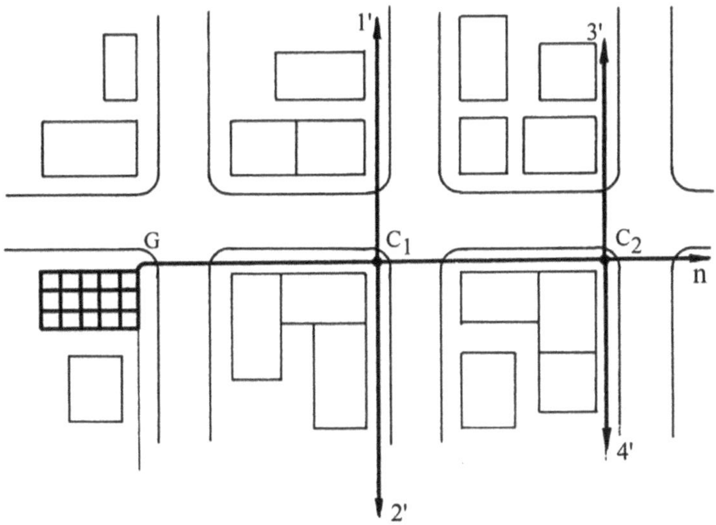

Figure 6.2 One primary and several possible secondary directions for laying cable lines [1].

If too many cable lines are laid in the same cable trench, their transmission capacity is lowered, so that, as a rule, each direction is used for approximately the same number of cables. The space disposition of cables that belong to a single primary direction can, in the general case, be represented by the situation shown in Fig. 6.2. This does not mean that all available directions will actually be used, especially at the time the station is put into operation, when many of the planned MV cable lines are not yet installed.

It is assumed here that the area under the sidewalk at the other side of the street is reserved for other kinds of underground installations. Certainly, the left and right turns from the primary direction, in the general case designated by n, are possible only at crossroads C_1, C_2, etc. These turns do not have to be strictly at right angles, as shown in Fig. 6.2; it is sufficient that the directions are significantly different.

Fig. 6.2 shows a typical disposition of substation ground grid G and outgoing cable lines in an urban area, where $1'$, $2'$, $3'$, etc. represent the available secondary directions for installing at least one cable line.

The areas where the cable lines are laid are of interest up to distances that can be considered as the limiting ones from the point of view of the cable line grounding effects. Thus the first question that needs to be answered is: how can one define and determine the area involved by the grounding system of an HV/MV distribution substation?

6.2.2 Urban Area Covered by a Grounding System

This area can be approximately determined on the basis of the value of active cable line length, or the length along which an outgoing MV cable line performs its grounding function, when a raised potential appears at its beginning, i.e., in the HV/MV substation. Also this length can be defined as a length for which the transfer/axiak impedance of the cable line metal sheath becomes practically infinitely large. The analytical expression for estimating this length is derived under the assumption that the grounding impedance of a cable line whose length is equal to its active length is only 5% larger than the grounding impedance of the same cable line, but assuming that its length is infinite. This cable line length is, according to Eq. (2.94), estimated by:

$$L_a(\varepsilon = 0.05) \approx \frac{3m}{\ln\left(1 + \sqrt{\dfrac{\pi m |Z_S|}{\rho \ln \frac{L_a}{\sqrt{h d_{sh}}}}}\right)}, \tag{6.1}$$

where L_a is the active cable line length (m), Z_S is the self-impedance of the cable sheath of the length of 1 m (Ω), ρ is the equivalent soil resistivity along and around the cable line (Ωm), h is the depth at which the cable is laid (m), d_{sh} is the outer diameter of the cable sheath (m), m is the length of 1 m, and ε is the relative error in the value of impedance $Z_g(L_a)$ defined as $|Z_{c\infty} - Z_g(L_a)|/Z_{c\infty}$, where $Z_g(L_a)$ denotes the grounding impedance of the cable line whose length is equal to the active cable line length.

As can be seen, this is a transcendental equation in which the only unknown quantity is the length L_a. Such equations may be solved graphically by representing each side of this equation as the separate functions of L_a or by doing a few iterative calculations. Here it is interesting to say that the active cable line lengths determined by using Eq. (6.1) are 27%–37% larger than those obtained by the earlier known semiempirical formula given in Ref. [1].

In relation to Eq. (6.1), something else should also be pointed out. In urban areas the equivalent soil resistivity along and around a cable line cannot be obtained by standard geological measurements. Performing these measurements is practically not possible because this area is already occupied by many urban objects, such as buildings, squares, streets, sidewalks, and different underground metal installations. As a result, we are forced to adopt one, roughly estimated, value based only on the main geological characteristics of the local soil structure and use it as a datum necessary for determining cable line active length.

When the cable line active length is known, the area covered by the spontaneously formed grounding system is also known because it is not difficult to imagine that this area is approximately defined as a circular plate with the center in the considered HV/MV substation, and with the radius equal to the active cable line length.

On the basis of the quantitative analysis performed by using Eq. (6.1), it follows that, even for very low values of the specific soil resistivity and standard cable constructions, the active cable line length is not less than 300 m. For such large lengths, the number of intersections along a street can be different and according to this the cable lines laid in the secondary directions can be different. The common characteristic of these cable lines is that they usually feed many MV/LV substations aligned in a row, on average 200 m apart, so that the total lengths of MV cable feeders is usually significantly greater than their active length.

6.3 GROUND GRID AND AN EXTERNAL ELECTRODE

6.3.1 Proximity Effect Between a Grid and an External Electrode

When the elements of a grounding system are close together they interfere with each other while performing the function of fault current dissipating into the surrounding earth. This effect, known as the proximity effect, causes an increase in the grounding system impedance, compared to the situation when the same grounding system elements are of great mutual distances and connected with an ideal conductor.

When the ground wires of overhead lines are external grounding electrodes, relatively simple and sufficiently accurate analytical expressions have been developed (Chapter 2). A favorable circumstance in this case is that due to the great spatial distance between the substation ground grid and the tower footing electrodes, the proximity effect is practically nonexistent. However, in the case of metal sheaths of cables having noninsulating outer sheaths, as well as for all other similar grounding electrodes in direct and continuous contact with the ground and passing through the potential field of the substation grounding grid, the situation is significantly different. The calculations necessary to include this effect are much more complicated, thus it was usual practice in the past to ignore this effect. This procedure gives sufficiently accurate values of the grounding impedance of the grounding system only when the ground grid is relatively small, as in the case of an MV/LV substation. However, in the case of an HV/MV substation, this approximate procedure leads to an inaccurate evaluation of the impedance of the substation grounding system, ground grid potential rise, and transferred potentials.

The proximity effect may be described in the following manner. At the substation ground grid the potential of the cable sheath, the grid, and the surrounding earth is the same. Because of the different conductivities of the soil and cable sheath, the voltage drops or voltage profiles in a radial direction for the substation ground grid differ in line with the distance from the grid increases. This voltage difference is not linear but reaches a maximum at some distance and then decreases until at a "remote ground" they are both effectively at zero potential. There exists therefore a very complex interrelationship that cannot be taken into account by simple mathematical models. First, the problem was solved by using complex numerical methods based on the application of modern computers, but even then, certain simplifying assumptions are made. However, here are presented the analytical expressions enabling approximation of the appreciation interference between a substation ground grid and an outgoing cable line with sufficient accuracy.

As in former analyses, consideration will start from the simplest case, i.e., when the grounding system is composed of a substation ground grid and only one, from the standpoint of the grounding effects, infinitely long cable line. Impedance of such a grounding system can be determined by determining separately the contribution of an external electrode (outgoing cable line) in dissipating the fault current into the earth. At this we assume that the dissipation of the fault current by the ground grid proceeds without any interference of the cable line, and that the total proximity effect is manifested only through the value of the current that is dissipated through the cable line sheath. Under such assumption, the ground grid resistance of an HV substation can be determined in many practical situations by Eq. (4.19).

In the case of indoor (conventional or gas-insulated) substations instead of a ground grid, the steel armature of the substation foundation acts as a substation grounding electrode. Then, we can substitute the area covered by the substation building with an equivalent circular plate, and for determining the resistance of such a grounding electrode use the following approximation:

$$R_g \approx \frac{\rho}{2D_e}, \ \Omega, \tag{6.2}$$

where D_e is the diameter of the equivalent circular plate (m).

The condition for the approximate equivalence of the armature of a substation foundation (or ground grid in the case of an outdoor substation) and the circular plate electrode is the equality of the covered area. If we denote this area as A the diameter of the equivalent circular plate is determined by:

$$D_e = \sqrt{\frac{4}{\pi}}\sqrt{A} = 1.128\sqrt{A}(\mathrm{m}). \tag{6.3}$$

The potential distribution in the vicinity of a grounding electrode, such as the grounding grid of HV installations, can be approximately described by the analytical expression derived for the case of a flat circular grounding electrode on the surface of the earth, as shown in Fig. 6.3.

The used notation has the following meaning:

$V_1, V_2, \ldots, V_x$—potentials along the foreseen path for installing a cable line,

I_0—current dissipated into the earth through the cable sheath (A),

R_S and Z_S—lumped parameters of the lumped parameter ladder circuit representing the cable line sheath (Ω),

G—substation ground grid,

r_e—radius of the equivalent circular plate electrode expressed in relation to the length of 1 m, and

$0'$—center of the equivalent circular plate electrode.

On the basis of the adopted approximation, the relative value of the potential at an arbitrary distance x from the center of the equivalent circular plate grounding electrode, of diameter D_e, is, according to Ref. [5], given by:

$$V_x = \frac{2}{\pi}V_0 \arcsin\frac{D_e}{2x}, \tag{6.4}$$

where V_0 is the potential of the substation grounding electrode (V) and x is the arbitrary distance along the cable line seen from $0'$ (Fig. 6.3) in meters. On the basis of the equivalent circuit in Fig. 6.3B it is obvious that these potentials are the only reason of the existence of the so called proximity effect.

However, if it is assumed that a cable line is already laid and that its metal sheath is connected to the grid, for an unchanged value of the total fault current, the potential on the grid would decrease and the earth potential distribution would be changed in the extend unknown for us. However, the only what we need is the value of the grounding impedance of such a grounding system. Therefore it will be assumed that the potential

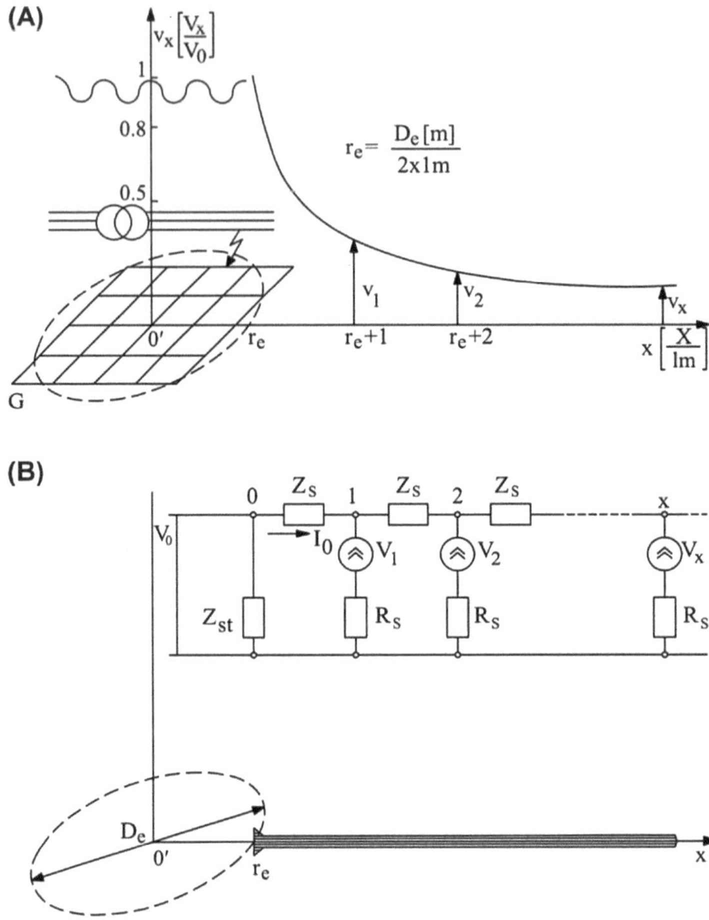

Figure 6.3 The substation grounding grid and an outgoing cable line [3]. (A) Surface potential profile around a ground grid only. (B) Cable line and its lumped parameters equivalent circuit.

distribution formed by dissipating the fault current through the ground grid remains unchanged after the cable line laying. Certainly, this assumption means that the value of the resistance of the ground grid will be the same as in the case when only the ground grid exists. Furthermore, under this assumption for a certain value of the potential of the grid, V_0, the potentials created by dissipation of the fault current at any point along the cable line can be considered as known ($V_x = v_x V_0$). Thus the whole proximity effect is expressed through these earth potentials represent individual voltage sources, as shown in the equivalent circuit of Fig. 6.3.

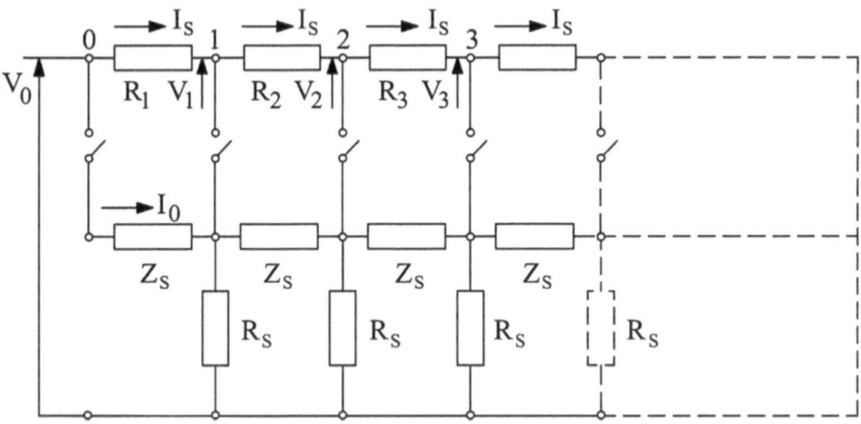

Figure 6.4 The equivalent circuit before establishing the current through the cable sheath [3].

In Fig. 6.3, R_1, R_2, … , R_n, … represent resistances of individual segments of the soil, which encompass an imaginary conic tube touching the soil surface along the cable line and formed under the assumption that an arbitrary chosen current I_S, dissipated from the ground grid through this tube, has the same value along its whole length, as shown in Fig. 6.4.

From the circuit of Fig. 6.4 it is clear that at the instant when the current flows are established through the cable sheath, current I_S and all the potentials this current creates on the earth along the cable line decrease.

However, for an unchanged value of voltage V_0, current I_0 diverted by the line metal sheath increases by the same amount, with respect to the value that this current would have had if current I_S and the resulting potentials along the cable line had remained unchanged. The proof is given by using the circuit of Fig. 6.5, which is a simplified version of the circuit of Fig. 6.4 obtained under the assumption that the cable is of finite length and that it can be represented by one discretized section only.

According to Fig. 6.5, the sum of the currents that are dissipated through the cable line sheath and the surrounding soil is:

$$I_0 + I_S = \frac{V_0(R_1 + Z_{sh})}{R_S R_1 + R_S Z_S + R_1 Z_S}. \tag{6.5}$$

However, if voltage V_1, i.e., the voltage that exists before the appearance of current I_0, is known, it follows that:

$$I_0 + I_S = \frac{V_0}{R_S + R_1} + \frac{(V_0 - V_1)(R_1 + Z_{sh})}{R_S R_1 + R_S Z_S + R_1 Z_S} + \frac{V_0 + V_1}{R_1}. \tag{6.6}$$

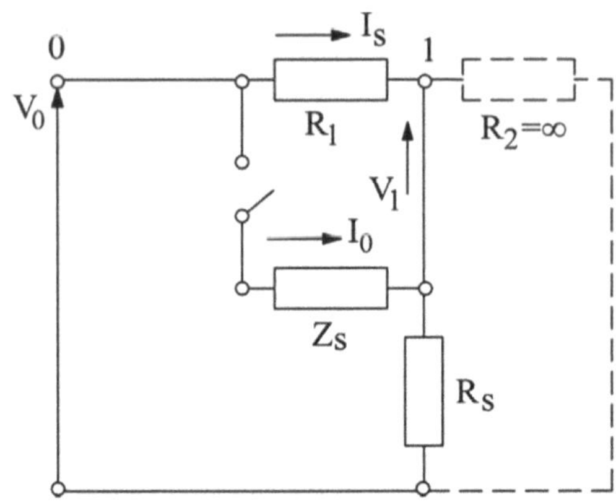

Figure 6.5 Simplified version of the circuit presented in Fig. 6.4 [3].

Since, according to the given definition, current I_S represents a part of the total current dissipated by the ground grid, it is clear that the total current dissipated by the grid will decrease due to the proximity effect if the value of voltage V_0 remains unchanged. However, the previous discussion served only to show that the total proximity effect could be involved if we have assumed that it appeared only in the cable line acting as the external grounding electrode. As soil resistivity is much larger than the impedance of the cable sheath for the same discretized section ($R_1 \gg |Z_S|$), it shall be assumed that impedance Z_S includes the influence of the soil, seen as a parallel branch between the corresponding two points along the cable line.

From the previous consideration, all circuit parameters and voltages in Fig. 6.3B are known and one can determine all currents in this circuit, including current I_0. This current represents a part of the fault current being dissipated by the metal sheath of the cable line under the assumption that only this current is reduced because of the proximity effect between the ground grid and the cable line as grounding conductor. The relationship between voltage V_0 and this current will be taken as the calculated/apparent grounding impedance of the cable line. However, this impedance expresses the real contribution of the cable line in dissipating the ground fault current, for it completely includes the mutual proximity effect with regard to the assumption that dissipation of the fault current through the ground grid is unhindered. By this, it is necessary to have in mind that in practical conditions the grounding contribution of the outgoing cable line are always accompanied by the proximity effect. Therefore through

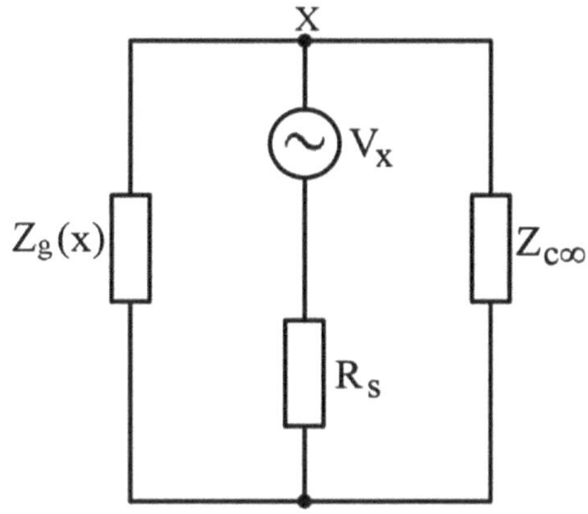

Figure 6.6 The simplified equivalent circuit seen from an arbitrary point, X [3].

this impedance the effect that proximity between the ground grid and the cable line has on dissipation of the fault current through the whole grounding system, or on impedance of the whole grounding system, is involved.

In accordance with the equivalent circuit in Fig. 8.3, the mentioned impedance can be determined by using the method of superposition. All currents appearing at the beginning of the cable line (the point of connection of the metal sheath and the ground grid), or at a distance from the center of the grid (point 0), which is equal to the equivalent radius, r_e, should be superposed. To do this, an infinite cable line equivalent circuit in which only voltage V_x is present is shown by the simplified equivalent circuit in Fig. 6.6.

In the circuit in Fig. 6.6, impedance $Z_g(x)$ represents the grounding impedance of the cable line seen from an arbitrary point, X, along the cable line toward its beginning. It is evident that the value of this impedance depends on the distance between this point and the beginning of the cable line. The only exception occurs when the ground grid resistance, R_g, connected at the beginning of the line is approximately equal to the grounding impedance of the infinite cable line, $Z_{c\infty}$. In that case grounding impedance of the cable line, seen from any point along the cable line toward its beginning, has a constant value, equal to the grounding impedance of the infinite cable line. Since in many cases we can adopt $R_g \approx Z_{c\infty}$ (Fig. 6.3B), it follows that $Z_g(x) = Z_{c\infty}$, which greatly simplifies further analysis. On the basis of this approximation, it follows that the potential at point X, when the currents through the cable sheath appear, is:

$$V'_x = V_x\left(1 - \frac{2R_S}{2R_S - Z_{c\infty}}\right), \tag{6.7}$$

where $Z_{c\infty}$ is the grounding impedance of the cable line when it can be treated as infinitely long.

Grounding impedance, $Z_{c\infty}$, is determined, according to Eq. (2.53), by:

$$Z_{c\infty} = \sqrt{R_S Z_S}. \tag{6.8}$$

The length of a segment of discretization is taken to be 1 m for practical purposes. Then, resistance of one discretized section of cable line metal sheath, according to, e.g., Ref. [2], is determined by:

$$R_S = R_S(L_a) = \frac{\rho}{\pi m}\ln\frac{L_a}{\sqrt{hd_{sh}}},\ \Omega, \tag{6.9}$$

where L_a is the active cable line length (m), d_{sh} is the external diameter of the cable sheath (m), h is the cable burial depth (m), and m is the length of 1 m.

Impedance Z_S represents the self-impedance of the cable metal sheath and is determined by Eq. (2.55). The value of this impedance is nearly constant for widely differing values of the specific soil resistivity, e.g., 20–200 Ωm. However, in determining impedance Z_S of cables having steel armors, it is necessary first to estimate the value of the current through the cable sheath by using an iterative procedure. Its value will determine whether and when it is necessary to increase the reactive part of this impedance by the cable steel armor effect (1.4 Ω/km). In cases where, because of a decrease in the current along the cable line sheath, values of this impedance at different sections of the cable line are different, one should choose the value that guarantees that the final results will be on the safe side. This means practically that one should choose its larger value in determining the grounding impedance of the whole grounding system, whereas the smaller value should be taken when determining the largest value of potential transferred through the cable line sheath.

According to Fig. 6.6, potential V_x can be taken as a voltage source whose internal resistance is equal to resistance $R_S(L_a)$. Then, on the basis of the expression for the equivalent transfer impedance for a certain section of an infinite cable line, when determining the current that creates potential V_x at the beginning of the cable line, one can use the circuit in Fig. 6.7.

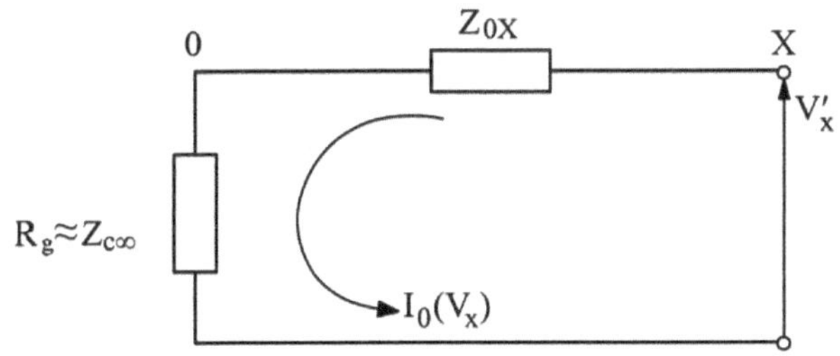

Figure 6.7 The equivalent circuit for determining the current $I_0(V_x)$ [3].

In the given equivalent circuit, Z_{0X} represents the equivalent axial impedance of the cable sheath between 0 and X, which is, according to Eq. (2.12), determined by:

$$Z_{0X}(Z_{st} \approx \;\; Z_{c\infty}) = \frac{k^x - 1}{k - 1} Z_S, \tag{6.10}$$

where k represents the current distribution factor along the metal sheath of an infinite cable line determined by:

$$k = 1 + \frac{Z_{c\infty}}{R_S}. \tag{6.11}$$

On the basis of the equivalent circuit in Fig. 6.7 and Eqs. (6.7) and (6.10), the current created by the voltage V_x at the beginning of the cable line is:

$$I_0(V_x) = \left(1 - \frac{2R_S}{2R_S - Z_{c\infty}}\right) \frac{V_x}{\dfrac{k^{x-r_e} - 1}{k - 1} Z_S + Z_{c\infty}}. \tag{6.12}$$

It is already mentioned that the length of the section of discretization is equal to 1 m. Then, by using the simple relation $x = r_e + n$, all the points of the equivalent circuit of Fig. 6.3B can be mapped onto the x axis, where distances are given per unit values. This makes it possible, by using the Eq. (6.4), to express voltage V_x as a function of the distance from $0'$ and to transform Eq. (6.12) into:

$$I_0(V_x) = \frac{2V_0}{2\pi R_S + \pi Z_{c\infty}} \frac{\arcsin \dfrac{r_e}{x}}{k^{x-r_e}}, \tag{6.13}$$

where r_e is the radius of the circular plate equivalent to the area covered by the substation ground grid:

$$r_e = \sqrt{\frac{A/m^2}{\pi}}. \tag{6.14}$$

Now, when the current that potential V_x creates at the beginning of the cable line is known, the current created by the earth potentials along the whole length of the cable line, according to Fig. 6.3, is:

$$I_0' = C_1 \int_r^{\infty} k^{-x} \arcsin\frac{r_e}{x} dx, \tag{6.15}$$

where:

$$C_1 = \frac{2V_0 k^{r_e}}{2\pi R_S + \pi Z_{c\infty}}. \tag{6.16}$$

By solving the definite integral one obtains:

$$I_0' = \frac{r_s}{1 - \ln k} \frac{V_0}{2R_S + Z_{c\infty}}. \tag{6.17}$$

In the electrical circuit shown in Fig. 6.3B there is also a voltage V_0 that acts directly through the conductive connection of the cable sheath with the ground grid. The current created at the beginning of the cable, when this voltage is alone, according to the circuits in Figs. 6.3B and 6.4, is:

$$I_0'' \frac{V_0'}{Z_{c\infty}}. \tag{6.18}$$

Since current I_0' is following in the opposite direction in comparison with I_0'', current at the beginning of the cable line is:

$$I_0 = I_0'' - I_0' = \frac{V_0}{Z_{c\infty}}\left(1 - \frac{r_e}{1 - \ln k} \frac{Z_{c\infty}}{2R_S + Z_{c\infty}}\right). \tag{6.19}$$

Finally, on the basis of the expression for current I_0, the expression for calculating (apparent) grounding impedance of the cable line, $Z_{c\infty}'$, is:

$$Z_{c\infty}' = \frac{V_0}{I_0} = Z_{c\infty}\left(1 + \frac{r_e}{1 - \ln k} \frac{Z_{c\infty}}{2R_S + Z_{c\infty}}\right)^{-1}. \tag{6.20}$$

Since the adopted discretized section is only 1 m, parameter k is almost equal to 1, and Eq. (6.20) can be substituted by:

$$Z_{c\infty}' \approx \left(1 - \frac{r_e Z_{c\infty}}{2R_S + Z_{c\infty}}\right)^{-1} Z_{c\infty}. \tag{6.21}$$

Impedance $Z_{c\infty}'$ represents the apparent, not the real, grounding impedance of the infinite cable line, because it incorporates not only the influence of the substation ground grid on dissipation of the current through the cable line, but also the influence of the cable line on dissipation of the current through the substation ground grid. On the basis of the former consideration the impedance of the considered grounding system is:

$$Z_e = \left(\frac{1}{R_g} + \frac{1}{Z'_{c\infty}} \right)^{-1}. \tag{6.22}$$

This expression is well known, but here it involves the proximity effect between the cable line and the grounding electrode, because $Z'_{c\infty}$ is determined by Eq. (6.21). This expression can also be presented in the following form:

$$Z'_{c\infty} = (1 + p)Z_{c\infty}, \tag{6.23}$$

where coefficient p expresses the influence of proximity between the grounding electrode (grid) and the cable line determined by:

$$p = \frac{r_e Z_{c\infty}}{2R_S - (r_e - 1)Z_{c\infty}}. \tag{6.24}$$

Certainly, the proximity effect is less and less pronounced for more distant sections of the cable line. The coefficient p of the cable section, which begins at an arbitrary distance, l, from the grounding electrode and ends at the end of the cable line, or at infinity, according to Ref. [1], is determined by:

$$p(l) = p_l = \frac{2Z_{c\infty}(C_2 + r_e)}{\pi k^l (2R_S + Z_{c\infty}) - 2Z_{c\infty}(C_2 + r_e C_3)}, \tag{6.25}$$

where the coefficients C_2 and C_3 are determined by:

$$C_2 = (r_e + 1)\arcsin\frac{r_e}{r_e + 1}, \tag{6.26}$$

$$C_3 = \ln\left[\frac{r_e + 1}{r_e} + \sqrt{\frac{l^2}{r_e^2} + \frac{2l}{r_e}}\right]. \tag{6.27}$$

The quantitative analysis presented in this Section 6.3.2, shows that, even under conditions where the proximity effect is very pronounced, coefficient p_l becomes negligible for sections that begin at distances greater than 40 m. This implies that turns more distant than 40 m from the primary direction (directions 1′, 2′, 3′, and 4′ in Fig. 6.2) do not significantly influence the value of coefficient p. Thus Eq. (6.21) can also be used if the cable is laid along any of these directions. In other words, it means that the primary direction need not be selected to be strictly radial.

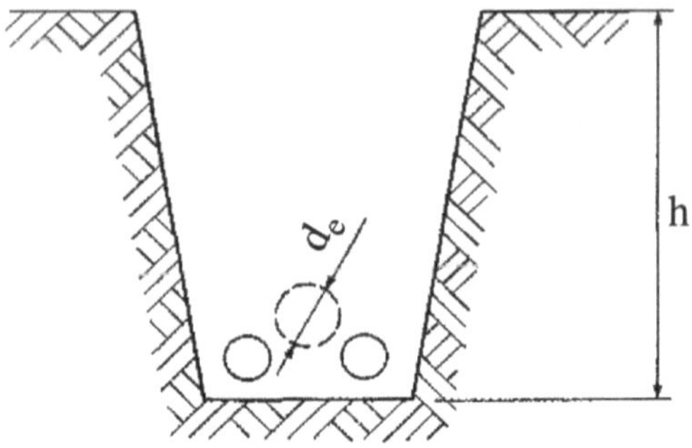

Figure 6.8 Two cables in the same trench.

The case when several cable lines are in the same trench along the whole of their active length can also be taken into consideration by substituting all these cables with only one fictitious cable having an equivalent external diameter. It can be determined that the surface along which contact with earth is established remains unchanged. This is illustrated by Fig. 6.8.

Because of the need for standardization, used cables usually have the same constructive parameters, so that the equivalent diameter of the fictitious cable is determined by multiplying cable diameters by the number of commonly laid cables. Subsequently, Eq. (6.21) can be used in these situations as well.

As the value of the external cable diameter has little influence on the value of the imaginary part of the self-impedance of the cable sheath, one can use the following approximation for determining this impedance for the equivalent cable:

$$Z_e \approx \frac{R_{sh}}{N_c} + jX_{sh}, \tag{6.28}$$

where R_{sh} (X_{sh}) is the real (imaginary) part of the self-impedance of the metal sheath of one cable (Ω) and N_c is the total number of cable lines in the common trench.

6.3.2 Method Verification

Impedance of the grounding system consisting of one copper ground grid and one cable line acting as a grounding conductor, as shown in Fig. 6.9, has been calculated in two ways: by using the methods that are based on computer application and by using the analytical expression derived here (Eq. 6.21).

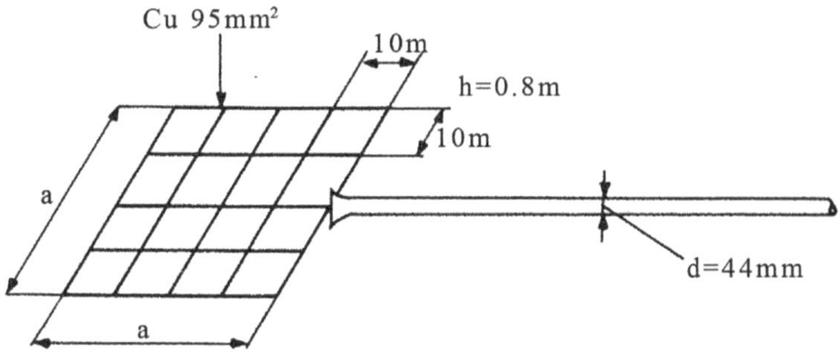

Figure 6.9 The considered grounding system [3].

Table 6.1 Results obtained using a computer [3]

ρ (Ωm)	a (m)	R_g (Ω)	$Z_{c\infty}$ (Ω)	R_e (Ω)	R'_e (Ω)	$(R'_e - R_e)/R'_e$ (%)
20	40	0.2281	0.2153	0.1106	0.1303	17.8
20	80	0.1185	0.2153	0.0764	0.0893	16.9
60	40	0.6843	0.6459	0.3318	0.3909	17.8
60	80	0.3555	0.6459	0.2292	0.2679	16.9

The results obtained by using a computer, for different values of the relevant parameters, are shown in Table 6.1.

In addition to the grounding resistance of the grounding system calculated without the proximity effect, R_e, and with this effect taken into consideration, R'_e, Table 6.1 gives the grounding resistances of the ground grid as well as the cable line, seen separately (R_g and $Z_{c\infty}$, respectively). The active cable line length in both considered cases is considered to be 300 m.

For the given values of the relevant parameters, the results obtained by using Eq. (6.21) are presented in Table 6.2. The results presented in Table 6.2 are obtained under the assumption that the impedance per unit length of the cable sheath is equal to (1.8 + j3.7) Ω/km, which corresponds to cases where the currents dissipated through the metal sheath exceed 150 A.

Table 6.2 Results obtained using Eq. (6.21) [3]

$Z_{c\infty}$ (Ω)	Z_e (Ω)	Z'_e (Ω)	$\lvert Z'_e \rvert$ (Ω)	$(\lvert Z'_e \rvert - \lvert Z_e \rvert)/\lvert Z'_e \rvert$ (%)
0.1778 + j0.1110	0.0771 + j0.0835	0.0610 + j0.1129	0.1283	13.0
0.1778 + j0.1110	0.0480 + j0.0624	0.0030 + j0.0937	0.0937	19.0
0.3074 + j0.1923	0.1797 + j0.1525	0.1714 + j0.2015	0.2646	7.7
0.3074 + j0.1923	0.1245 + j0.1394	0.0887 + j0.1944	0.2137	14.7

The influence of the proximity effect, expressed in percent, is somewhat larger for the results given in Table 6.1, but it should be noted that the results presented in Table 6.1 are not absolutely correct, since the method relying on computer calculations is not developed without approximations. As can be seen, the assumptions and simplification made in obtaining Eq. (6.21) are not of such magnitude that they have a significant influence on overall accuracy and applicability in solving practical problems.

Then, it has been assumed that the grounding system shown in Fig. 6.9 had the following relevant data: $a = 100$ m, the impedance per unit length of the cable sheath $Z'_{sh} = (0.7 + j0.7)$ Ω/km, which corresponds to cases where the current dissipated through the cable exceeds 500 A, and the equivalent soil resistivity is 20 Ωm. On the basis of the given data, we will consider the case where the proximity effect is very pronounced. By using Eq. (6.25), change in the value of coefficient p_l for the section of a cable ending at infinity and with the beginning of the distance l from the ground grid constantly increasing, is shown in Fig. 6.10.

The analytical expression (6.21), with slight approximations, can take into account the proximity effect between the ground grid and long external grounding electrode, such as the cable line with a noninsulating outer sheath. The use of these expressions in solving practical problems enables a sufficiently correct estimation of the contribution of a cable line as a grounding conductor in the framework of the grounding system consisting of one ground grid and one cable line acting as grounding conductor (Fig. 6.9).

The considered proximity effect is so pronounced that, for security reasons, it should not be neglected in estimating the safety conditions of HV substations. The possibility of taking into consideration the proximity effects between grounding system elements is especially important for

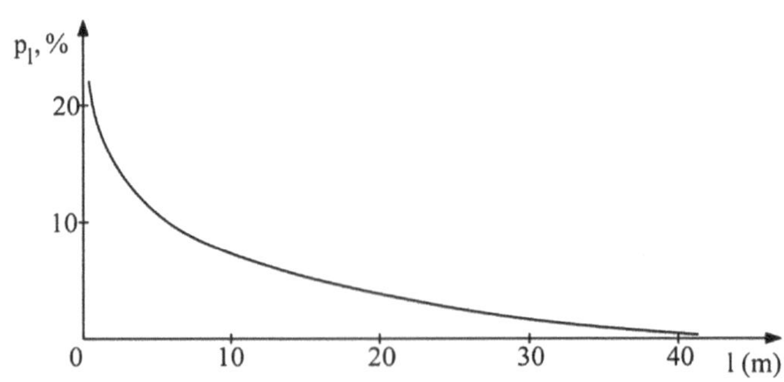

Figure 6.10 Change in the value of coefficient p_l along the cable line [1].

grounding systems formed in typically urban conditions, i.e., where there are many cable lines and where their grounding contribution is very important.

6.4 GROUNDING SYSTEM FORMED OF MORE CABLE LINES

6.4.1 Grounding Contribution of Cable Lines Laid in Secondary Directions

In this case, let us assume one more cable line in comparison with the previously considered case. Under the assumption that this line is laid in one of the secondary directions (Fig. 6.2) the physical model and the corresponding electrical circuit are presented in Fig. 6.11.

The used notation has the following meaning:

G—ground grid of the HV/MV substation,

V_G—potential of the substation ground grid,

C—place of cable line divergence from a certain primary direction,

Z_{GC}—axial impedance of the cable sheath on the common section, *GC* (Ω),

$Z_{c\infty}$—grounding impedance of the cable line seen from point *C* toward the end of the cable line (Ω),

L—relative length of the common section, *GC*, in which the cable line is laid together with the other cable line belonging to the same primary direction, and

n (*m*)—primary (secondary) direction in which at least one cable line is installed.

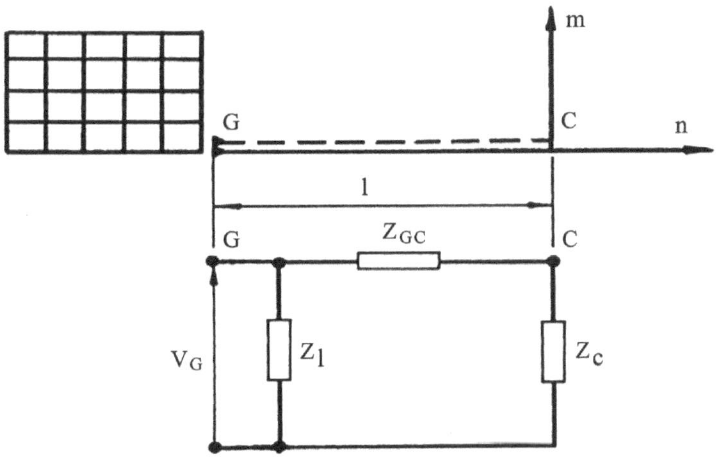

Figure 6.11 Cable line laid in one of the secondary directions [1].

Grounding impedance of the cable line laid in the primary direction is determined using Eq. (6.21) as if the cable line in the secondary direction did not exist. In this way, it is approximated that the cables are laid side by side along the section from the ground grid to the street intersection (from G to C in Fig. 6.11), which decreases the fault current dissipation capabilities of both cables. The fact that they interfere with each other and their mutual inductive coupling along this section is taken into consideration by increasing only the value of the grounding impedance of the cable along the secondary direction. This is done by completely neglecting the current dissipated to earth by this cable line along the whole section where they are laid side by side, i.e., by treating impedance Z_l (Fig. 6.11) as infinitely large. Under this assumption, by using the equivalent circuit shown in Fig. 6.11, the grounding contribution of the cable laid along the secondary direction, which is, in general, designated by Z_m, is:

$$Z_m = Z_{GC} + Z_{c\infty}. \tag{6.29}$$

When the cable line can be treated as infinite, the equivalent axial impedance Z_{GC}, according to Eq. (2.83), is determined by:

$$Z_{GC} = (k^l - 1)Z_{c\infty}. \tag{6.30}$$

According to Eqs. (6.29) and (6.30) the grounding contribution of the cable line laid in any (mth) secondary direction is determined by:

$$Z_m = k^l Z_{c\infty}. \tag{6.31}$$

In cases where length l is equal to or larger than the active cable line length ($l \geq L_a$), axial impedance Z_{GC} becomes practically infinitely large and the grounding contribution of such cable lines can be completely disregarded as cable lines laid in secondary directions.

Also, in some practical situations, the nearest MV/LV substation is just behind the place of the cable line divergence and has impedance of its own grounding system that is significantly lower than impedance $Z_{c\infty}$. Such a situation is schematically presented in Fig. 6.12.

The used notation has the following meaning:

L—relative length of the cable line from C to the nearest supplied MV/LV substation expressed in relation to the length of 1 m.

In such cases, instead of Eq. (6.30) it is more correct to use the following expression:

$$Z_m \approx (k^l - 1)Z_{c\infty} + Z_C. \tag{6.32}$$

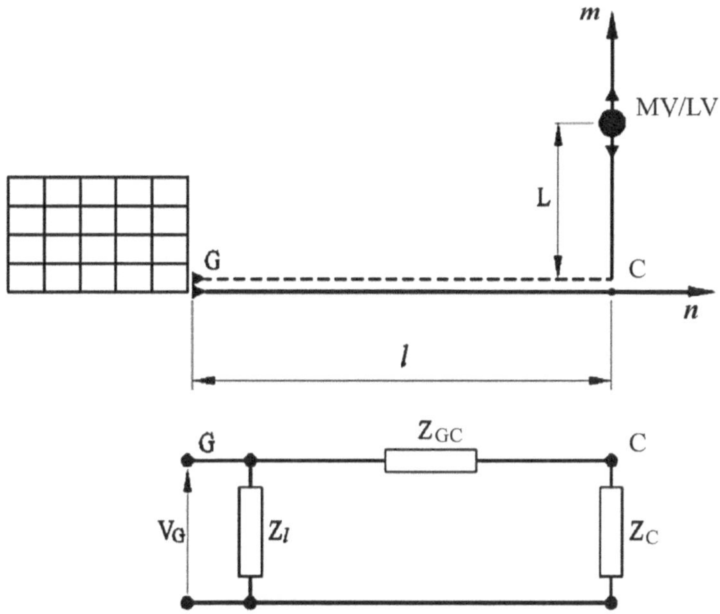

Figure 6.12 Cable line laid in one of the secondary directions and a nearby MV/LV substation [2].

On the basis of Eq. (2.67) impedance Z_C is determined by:

$$Z_C = \frac{(k^L - k^{-L})Z_{c\infty} + (k^L + k^{-L})Z_{gs}}{(k^L + k^{-L})Z_{c\infty} + (k^L - k^{-L})Z_{gs}} Z_{c\infty}, \tag{6.33}$$

where Z_{gs} is the grounding system impedance of the nearest supplied MV/LV substation excluding the grounding effect of the feeding section of the considered cable line.

The value of impedance Z_{gs} is usually obtained by on-site measurements. Thus through the value of this impedance, the grounding effects of many local underground metal installations, connected through LV neutrals and typical for urban areas, can be taken into account.

When impedance Z_{gs} is considerably smaller than impedance $Z_{c\infty}$ ($Z_{gs} \ll Z_{c\infty}$) and length L is no longer than 200 m, the following simpler expression can be used:

$$Z_C \approx \frac{(k^L - k^{-L})}{k^L + k^{-L}} Z_{c\infty}. \tag{6.34}$$

6.4.2 Impedance of the Whole Grounding System

The previous considerations and derived expressions enable one to obtain the grounding contribution of all cable lines belonging to an HV power installation by using the following expression relevant for grounding systems with mutually distant components:

$$Z_{eq} = \left(\sum_{n=1}^{N} \frac{1}{Z_n} + \sum_{m=1}^{M} \frac{1}{Z_m} \right)^{-1}, \tag{6.35}$$

where Z_{eq} is the equivalent impedance expressing the grounding contribution of all MV cable lines belonging to a considered HV substation (Ω), Z_n is the impedance expressing the grounding contribution of the cable line laid in an arbitrary (*n*th) direction belonging to the primary directions (Ω), Z_m is the impedance expressing the grounding contribution of the cable line laid in an arbitrary (*m*th) direction belonging to the secondary directions (Ω), N is the total number of cable lines laid in the primary directions, and M is the total number of cable lines laid in the secondary directions.

Finally, the impedance of the whole grounding system of an HV distribution substation is:

$$Z_{gs} = \left(\frac{1}{R_g} + \frac{1}{Z_{eq}} \right)^{-1}. \tag{6.36}$$

As can be seen, Eqs. (6.31) and (6.32) are obtained under the assumption that cables belonging to different secondary directions are disposed in such a manner, and so far apart, that their mutual interference in dissipating the fault current into the earth can be neglected (Fig. 6.2). This approximation, as is well known, is in accordance with almost all situations that can be encountered in practice.

The given expressions correspond to the situation where each of the primary directions is used for laying at least one cable line. If this is not the case, then it is necessary to choose the direction with the most distant turns from the imagined primary direction and treat it as the primary one. Consequently, all cables that turn in respect to it are then treated as cable lines laid in the secondary directions.

In deriving the given expressions, grounding effect (dissipating a fault current into the surrounding earth) of the cable lines that run in the secondary directions on the sections that coincide with the primary direction have been neglected. In this way, the interference of the grounding electrode in the dissipation of the fault current from these sections is also included.

Finally, by using Eqs. (6.32–6.34), one can take into consideration the useful influence of the cable line sheath as the electrical connection with the grounding systems of all nearby consumer (MV/LV) substations. Through the grounded secondary neutrals of these substations, even very distant underground metallic structures become incorporated into the grounding system of an HV distribution substation.

6.5 GROUNDING SYSTEMS HAVING CABLE LINES WITH COVERED METAL SHEATHS

6.5.1 The Calculation Procedure

The reasons why modern extruded dielectric cables, and especially cables with cross-linked polyethylene (XLPE), are increasingly replacing the so-called classical or paper-insulated cables are better exploitation characteristics, easier maintenance, and more acceptable prices. Bearing in mind this clearly marked trend, it is not difficult to conclude that different types of modern XLPE cables will be mostly used in distribution MV networks in the foreseeable future. As these cables are manufactured with an insulation layer over their metal sheath, they do not act as grounding conductors, i.e., they do not dissipate ground fault current directly into the earth. As a result, a question has been raised: how will this change in the type of utilized MV cables reflect upon the possibilities of solving of the grounding problem of distribution substations, especially those where this problem is the most pronounced—the HV/MV distribution substations?

In the national standards of certain European countries the thermoplastic oversheath was strictly required for every kind of MV cable to prevent corrosion of the metallic sheath directly in contact with the soil. However, in the contemporary power distribution networks the MV cable lines with uncoated metal sheaths will still be used. Due to this, a very spacious and complex grounding system is formed in a wide urban area around distribution HV/MV substations. The extremely favorable characteristics of such grounding systems enable a safe and cost-effective solution to the grounding problem of HV/MV substations even when total ground fault currents reach very high values, over 20 kA. Thus the question is: what can be expected if these cables are completely replaced by cables having insulated metal sheaths/screens? This question can be expressed in an even more direct way: will a widespread use of cables having insulated sheaths/screens significantly deteriorate general conditions for solving the grounding problem of HV distribution substations? Since the answer to this question was, at first sight, affirmative, many distribution companies adopted the practice of laying copper wire, or galvanized steel strips, in the trench together with cable lines having insulated sheaths/screens. This is obviously done with the intention of compensating for the absence of the grounding function of the cables having insulated metal sheaths/screens.

The analysis that will be presented here shows that the property that was, until recently, treated as a shortcoming can in urban conditions have certain advantages over cables having insulated metal sheaths/screens. Through the insulated metal sheaths acting as a solid conductive connection between grounding systems of MV/LV substations, a significantly larger grounding system is formed than that formed through cables having uncovered sheaths. Certainly, the obvious advantage is that unfavorable proximity effects among the main elements of the grounding system do not exist.

First, let us consider only a single MV cable line that supplies in series a certain number of MV/LV substations. By using the usual approximations and idealizations of the real electric circuit formed through the metal sheaths/screens of three single-core cables and the grounding electrodes of the supplied substations, such a cable line can be represented by the equivalent circuit shown in Fig. 6.13.

The used notation has the following meaning:

Z_c—grounding (input) impedance at the beginning of the cable line (Ω),

Z_{sh3}—self-impedance of three parallel metallic sheaths of the line section equal to the average length of all sections between two adjacent customer substations (Ω),

Z_{av}—average value of the grounding system impedances of all MV/LV substations without the grounding effects of the considered MV cable line (Ω),

Z_b—grounding impedance of the HV substation located at the end of the MV cable line (Ω), and

N—total number of substations connected to the considered cable line.

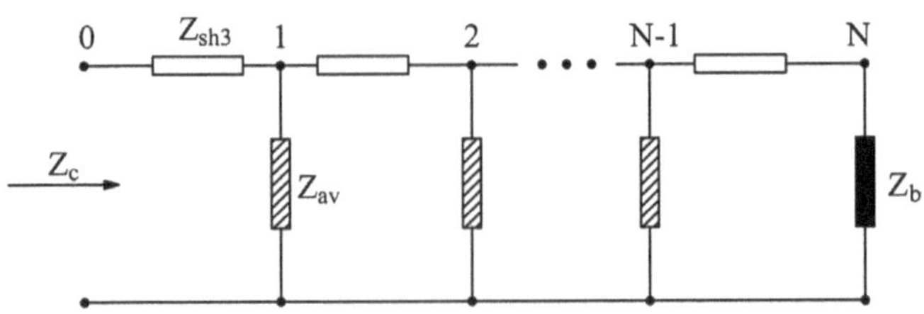

Figure 6.13 The equivalent circuit of a cable line and its grounding connections [2].

The idealized equivalent circuit shown in Fig. 6.13 obviously represents a uniform discrete parameter ladder circuit. By using Eq. (2.69), and having in mind that the grounding impedance of an HV distribution substation at the end of the line is usually very small ($Z_b \in [0.05, 0.2]\ \Omega$), impedance Z_c can be determined by:

$$Z_c(Z_b \approx 0) \approx \frac{k^N - k^{-N}}{k^N + k^{-N+1}} Z_{c\infty}. \tag{6.37}$$

In this expression k represents the coefficient of current distribution in circuit nodes (1, 2, 3, … , $N - 1$), under the assumption that the number of nodes is infinite ($N \rightarrow \infty$). It is, in accordance with Eq. (2.6), determined by:

$$k = 1 + \frac{Z_{c\infty}}{Z_{av}}, \tag{6.38}$$

where impedance $Z_{c\infty}$ represents the input impedance of the equivalent circuit when the number of nodes can be treated as infinite ($N \rightarrow \infty$). This impedance is, in accordance with Eq. (2.59), given by:

$$Z_c(N \rightarrow \infty) = Z_{c\infty} = \frac{Z_{sh3}}{2} + \sqrt{Z_{av} Z_{sh3} + \frac{Z_{sh3}^2}{4}}. \tag{6.39}$$

Impedance Z_{sh3} can be approximately determined by using Eq. (6.28). If cable lines are equal to or longer than their active length, Eq. (6.39) can be used instead of Eq. (6.36). When the section to the nearest MV/LV substation is significantly different from the average section between two adjacent consumer substations, it is not difficult to see on the basis of the equivalent circuit in Fig. 6.13 that $Z_{c\infty}$ is:

$$Z_{c\infty} = Z_0 - \frac{Z_{sh3}}{2} + Z_{c\infty} = Z_0 + \sqrt{Z_{av} Z_{sh3} + \frac{Z_{sh3}^2}{4}}, \tag{6.40}$$

or, in the general case, according to Eq. (2.69), we have:

$$Z_c = Z_0 - Z_{sh3} + \frac{(k^N + k^{-N+1}) Z_N Z_{c\infty} + (k^N - k^{-N}) Z_{c\infty}^2}{(k^N - k^{-N+2}) Z_N + (k^N + k^{-N+1}) Z_{c\infty}}, \tag{6.41}$$

where Z_0 is the self-impedance of the cable sheaths of the first section of the cable line and Z_N is the grounding impedance at the end of the cable line or its section.

On the basis of the analogy between the electrical circuit of Fig. 6.13 and the circuit formed by the grounding wire(s) and tower footings of an overhead line, the active length of the considered cable lines can be, according to Eq. (2.80), estimated by:

$$L_a(\varepsilon = 0.05) \approx \frac{3}{\ln\left(1 + \sqrt{\frac{|Z_{sh3}|}{|Z_{av}|}}\right)}, \tag{6.42}$$

where L_a is the active cable line length expressed through the number of average cable line sections, ε is the relative error of the value of grounding impedance $Z_c(L_a)$, $(|Z_{c\infty} - Z_c(L_a)|/|Z_{c\infty}|)$, and $Z_c(L_a)$ is the grounding impedance of the cable line when its length is equal to its active length.

The given expressions are applicable if the following criterion is satisfied: there is no mutual interference between grounding systems of the supplied MV/LV substations in dissipating ground fault current into the surrounding earth. In practice, these substations are located at a mutual distance, usually more than 100 m, so that the mentioned criterion is almost always satisfied. This is also valid for substations that are supplied from the same HV/MV substation, but through two different outgoing cable lines forming a closed loop. Certainly, this case should be considered, from the standpoint of grounding effects, as two different outgoing lines ending in the most distant of supplied MV/LV substations.

The certain deviation in relation to the equivalent circuit in Fig. 6.13 exists only in the proximity of an HV distribution substation. Because of the lack of space, the outgoing MV cable lines are laid in common trenches whose number usually cannot be greater than four. Due to this, the cable lines belonging to the same primary direction in the vicinity of an HV substation are disposed as shown in Fig. 6.14.

The used notation has the following meaning:

n—one of the primary directions of laying cable lines, $n \in [1,4]$, and
1 and 2—two secondary directions used for installing cable lines.

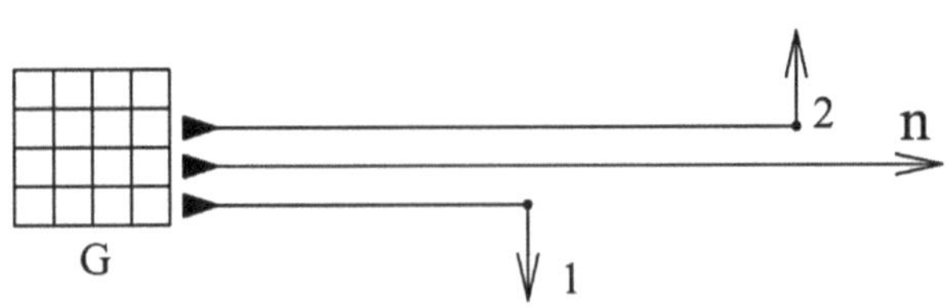

Figure 6.14 Cable line disposition in the vicinity of an HV substation [2].

When cable lines are laid in a common trench, as is the case of cable lines laid in any of the primary directions (Fig. 6.14), it is necessary to take into consideration their mutually inductive influence. However, a quite correct calculation in these cases is too complicated. Alternatively, it is possible that an approximate but sufficiently accurate calculation procedure can be described in the following manner. First, by using Eq. (6.40) one can calculate the grounding impedance of the cable lines seen from point 2 toward their ends (Fig. 6.14). Then, by treating their equivalent grounding impedance as an outgoing one and by assuming that between points 2 and 1 only one cable line exists, grounding impedance seen from point 1 toward the ends of the cable lines is calculated by using Eq. (6.41) if there is only 1 MV/LV substation between 1 and 2. Finally, by treating this impedance as an outgoing one and by assuming that between points 1 and *G* only one cable line exists, grounding impedance of all the cable lines belonging to a certain primary direction can also be estimated by using Eq. (6.41).

Certainly, the foregoing procedure can also be used for some other case, different from the standpoint of the number of cable lines belonging to the same primary direction. Finally, the grounding contribution of all outgoing cable lines can be determined as a parallel connection of the grounding impedances obtained separately for each of the primary cable line directions. By using the described calculation procedure, the complicated calculations of the mutually inductive influence on the common sections of cable lines in transferring the fault current through their cable sheaths are avoided. The described approximations of real electrical circuits ensure that the final results are slightly on the conservative side.

6.6 TRANSFERRED POTENTIALS

6.6.1 Cable Line Having Uncovered Metal Sheath

Very favorable characteristics of the considered grounding systems are accompanied by one undesirable effect known as the transfer of potentials. As a result, one more aspect of the cable lines acting as grounding conductors should be considered.

When the impedance of the grounding system and the maximum anticipated earth current, I_e, are known, one can determine the maximum anticipated potential of the substation ground electrode as a product of these two quantities. In cases where this potential exceeds the maximum permitted touch voltage, the relative distribution of potential on the earth surface above the ground grid becomes relevant. When this distribution is known, in general, one can determine the critical (highest) "touch voltage" and the "step voltage" that can appear in the yard covered by an HV substation. Determination of the critical touch voltages above the ground grid is possible by using the simplified procedure presented in e.g., Ref. [21], whereas in the case of gas-insulated substations it represents a specific problem that can be solved also by using Ref. [21]. However, for the grounding systems under consideration the highest potential differences (touch voltages) appear as a rule at a priori known places. These places are the nearest supplied MV/LV substations and their customer installations. This can be explained by the fact that large potentials are transferred to the zones that, because of their great distance from the substation grounding electrode (the source of the ground potential rise), have a very low, practically equal to zero, earth surface potential. Certainly, these conditions lead to the highest possible potential differences that can appear in the whole grounding system and their values reach almost the full value of the ground potential rise [21]. From this place, high potential is transferred to the places that are exposed to the general public and where the maximal tolerable touch voltages as a result are lower than those in the considered HV substation. Thus it is clear that the most critical potential differences that can appear in the whole grounding system of an HV substation located in urban or suburban areas are caused by transferred potentials. Certainly, in accordance with this there is no need for complicated surface potential calculations with the aim of estimating safety conditions for this type of HV substation.

Let us consider the simplest case where only one MV cable line is laid, feeding a number of consumer substations lying in a row and at approximately equal distances. In that case the equivalent circuit necessary for determining the transferred potential to the nearest substation is presented in Fig. 6.15.

The used notation has the following meaning:

S—the nearest supplied MV/LV substation,

V_G—maximal anticipated potential of the HV substation ground grid (V),

V_t—potential transferred to nearest substation *S*,

Z_{GS}—transfer/axial impedance of the cable sheath on the common section, from *G* to *S* (Ω),

Z'_S—grounding impedance of substation *S*, which does not include the grounding effect of the considered cable line in the section from *G* to *S*, and

l_s—length of the cable up to the nearest substation *S*, expressed in relation to the length of 1 m.

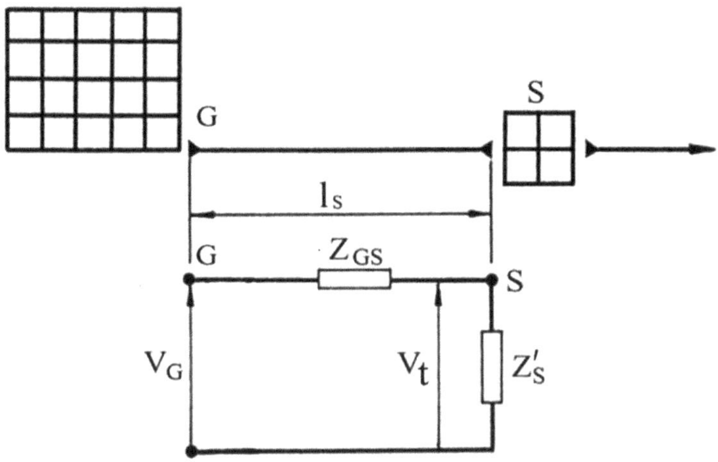

Figure 6.15 The equivalent circuit for necessary determining transferred potential [1].

If it is conservatively assumed that substation S does not have its own grounding electrode and grounding connections through the LV neutrals, according to Eq. (2.86), the value of the transferred potential can be estimated by approximation:

$$V_t(Z_{st} \approx Z_{c\infty}) \approx V_G\left(1 + \frac{Z_{c\infty}}{R_S}\right)^{-l_S} = V_G k^{-l_S}. \tag{6.43}$$

The potential V_G is determined by:

$$V_G = Z_{gs} I_e, \tag{6.44}$$

where I_e is the maximal anticipated earth current (A) and Z_{gs} is the grounding system impedance of the supply HV/MV substation, without taking into account the grounding contribution of the feeding section of the considered cable line (Ω).

Eq. (6.43) corresponds to the situation where impedance of the nearest substation, Z'_S, is approximately equal to impedance $Z_{c\infty}$. However, as the substations usually have their own grounding systems, which incorporate more than just the cable line under consideration, impedance Z'_S is usually smaller than impedance $Z_{c\infty}$. Therefore the results obtained by using Eq. (6.43) are usually too conservative. In critical situations, when a greater accuracy is desirable, on the basis of the equivalent circuit in Fig. 6.15 one can apply the following relation:

$$V_t = \frac{Z'_S}{Z_{GS} + Z'_{c\infty}} V_G. \tag{6.45}$$

In the given expression, Z_{GS} designates the equivalent axial impedance for the section of the cable line to the nearest MV/LV substation. For a cable line of known relevant parameters and a soil of estimated equivalent resistivity, this impedance would have the smallest value for Z'_S equal to zero. Then, according to Eq. (2.68), for an estimation erring on the side of caution:

$$Z_{GS}(Z'_S = 0) = \frac{k^{l_S} - k^{-l_S}}{2} Z_{c\infty}. \tag{6.46}$$

Since this is never the case in practice ($0 < Z_{s1} < Z_{c\infty}$), the calculated value of impedance Z_{GS} is always somewhat less than the actual one. Also, Eq. (6.46) simplifies the calculation of the transferred potential, making it unnecessary to know the actual value of impedance Z'_S and gives the results that are somewhat on the safe side.

When the cable is laid in the same trench with other cable(s), the analytical procedure for determining the transferred potential becomes much more complex. However, here one can make the extremely unfavorable assumption that the specific soil resistivity is infinitely large ($\rho \to \infty$). Then, according to Eq. (6.37), one obtains $R_S \to \infty$ and $k \to 1$, and the expression for the transfer impedance becomes:

$$Z_{GS}(\rho = \infty) = \lim_{k \to 1} Z_{GS} = \frac{d(k^{l_S} - 1)/dk}{d(k - 1)/dk} Z_S = l_S Z_S. \tag{6.47}$$

This expression gives the minimum possible value of the transfer/axial impedance of the cable sheath section if the cable had an isolated outer sheath. This value of the transfer impedance corresponds to the upper limit of the transferred potential determined by Eq. (6.45) and contains a large reserve from the standpoint of safety conditions.

When a more accurate evaluation is desired, influence of the proximity of other cable lines can be considered by introducing an equivalent cable line sheath substituting all existing ones (Eq. 6.28) and applying Eqs. (6.45) and (6.46).

The potentials transferred to the nearest MV/LV substation by earthed LV neutrals are further transferred to a large number of exposed metal structures in the LV network, such as cable connection boxes, lampposts, electrical home appliances, including all other underground metal installations typical for urban areas. Naturally, over a large urban area it is not possible to ensure that a certain coincidence of unfavorable circumstances will not cause the appearance of the full value of transferred potential exposed to human touch.

According to the previous conclusion, the transferred potentials become important in cases where the potential higher than permissible touch voltage can appear on the grounding electrode of the nearest MV/LV substation. Thus the following criterion should be satisfied:

$$V_t = k_t Z_{gs} I_e \leq V_p, \tag{6.48}$$

where V_t is the potential transferred to the nearest MV/LV substation (V), V_p is the permissible potential transferred to the nearest MV/LV substation (V), I_e is the maximal anticipated earth current (A), Z_{gs} is the impedance of the grounding system of the considered HV/MV substation (Ω), and k_t is the coefficient representing relative value of the potential transferred to the nearest MV/LV substation.

Depending on the type of earthing system implemented in an LV network, potential V_p can be somewhat higher than the maximal permissible touch voltage [22]. In cases where the nearest supplied MV/LV substation is not too close, metal sheaths of MV paper-insulated cables obtain axial impedances of high values as shown in Fig. 2.18 and significantly reduce the transferred potentials. However, this is not the case with cables having insulated metallic sheaths, progressively being more applied in contemporary power distribution networks.

6.6.2 Cable Line Having Covered Metal Sheaths

In the case of cables having insulated sheaths/screens, it is not difficult to show that coefficient k_t is determined by:

$$k_t = 1 - \frac{l_s Z_{sh3}}{Z'_S} \tag{6.49}$$

where l_s is the length of the cable line up to the nearest MV/LV substation (m), Z_{sh3} is the self-impedance of the sheaths of the three single-core cables per 1 m (Ω/m), and Z'_S is the grounding impedance of the nearest MV/LV substation, excluding grounding effect of the considered cable line as a connection with the grounding system of the supply HV substation (Ω).

On the basis of the presented analytical expressions and given the relevant data about the considered types of cables, it is not difficult to see that the value of coefficient k_t, for the same cable line length, is greater in the case of cables having insulated sheaths/screens. In other words, this means that the favorable transfer of the fault current to the distant grounding system elements is accompanied by undesirable better transfer of the high ground grid potential. At the same time, it means that the maximal earth current I_e at the same value of transferred voltage V_t can be greater in the case of cables having uncovered sheaths. However, as the nearest MV/LV substation is frequently very close to the HV/MV substation, usually at a distance of about 100 m, this advantage of cables having uncovered sheaths is not of greater practical significance.

Certainly, an analogous procedure for determining k_t can be applied in the case of other utilities such as gas pipelines, railway lines, or any other metal connection. Here it must be noted that voltages transferred along any of these metallic connections may be hazardous if a sufficient grounding electrode is not presented where these connections are earthed. At these installations too high transferred potentials can be prevented by using a corresponding isolating insert.

In cases of difficult grounding conditions when the criterion defined by Eq. (6.48) or (6.49) is not satisfied, the problem can be overcome by a general reduction of all potentials appearing on the HV substation grounding system. This can be done by applying the measures described in Section 4.3, or by increasing the number of outgoing MV cable lines, at the critical moment of putting newly built HV substations into operation. The additional measures leading to the reduction of only the substation grounding electrode resistance usually cannot give any significant effect, because the relative grounding contribution of the whole substation grounding electrode in such grounding systems is very small, especially in the cases with high soil resistivity.

6.7 QUANTITATIVE ANALYSIS

6.7.1 Relevant Data

The presented quantitative analysis is achieved by using a practical example of a 110/10 kV substation in the distribution network of Beograd, Serbia. It is a gas-insulated substation in surroundings that may be considered typical from the standpoint of urban conditions. The space disposition of this substation and the outgoing cable lines are shown on the diagram presented in Fig. 6.16.

The adopted primary directions of installing the cable lines are denoted by numbers 1, 2, 3, and 4.

The displayed urban area has the following dimensions: $(1400 \cdot 1400)$ m^2, while the foundation of the substation building covers an area equal to $(26 \cdot 52)$ m^2. The primary directions are adopted on the basis of the following criteria: the spatial disposition of two neighboring primary directions is mutually approximately perpendicular.

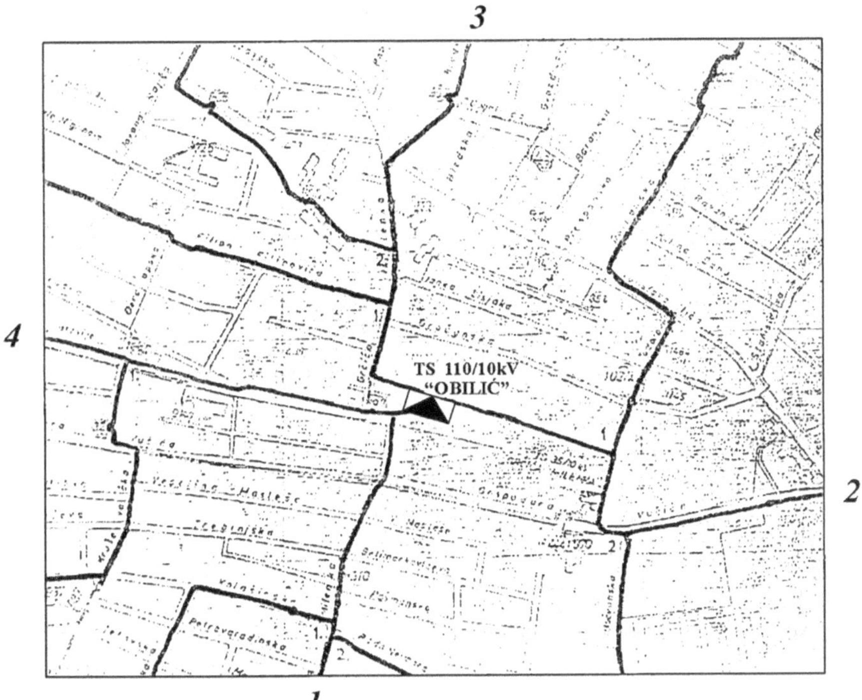

Figure 6.16 Disposition of the considered HV/MV substation and outgoing cable lines [2].

In one of the considered cases we will assume that all outgoing cable lines have uncovered metal sheaths, whereas in the other case we will assume that they have covered metal sheaths. The relevant data regarding the surrounding 10 kV cable network in the case of cables having uncovered metal sheaths are as follows:

- The total number of outgoing cable lines is 11.
- The places of divergences of the cable lines from the denoted primary directions are at the following distances seen from the HV substation building: for primary direction 1: 210 and 320 m; for primary direction 2: 300 and 390 m; for primary direction 3: 220 and 290 m; and for primary direction 4: 400 m.
- Cables having uncoated metal sheaths have an outer diameter of 44 mm and are buried at a depth of 0.7 m.
- The self-impedance of the metal sheath of the cables, obtained by measurements, is (0.7 + j2) Ω/km.
- The whole grounding system is located in an area with an equivalent soil resistivity estimated at about 30 Ωm.

- The influence of the conductive connections to the grounding system of the nearest supplied MV/LV substation in the case of cables having uncovered sheaths is completely disregarded.

The relevant data in the case of cables having insulated metal sheaths are:

- Total length of each of the MV cable lines is 4 km and each of them supplies nine 10/0.4 kV substations in series.
- Average value of the grounding impedance of all supplied substations $|Z_{av}|$ is 1.0 Ω.

Impedance Z_{sh3} is determined on the basis of Eqs. (2.55) and (6.28), as well as assuming that the equivalent current penetration depth is because of urban conditions (surrounding metal installations) equal to only 10 m (in accordance with the results of the experimental investigations performed in a typically urban area). Then, the value of this impedance is (0.100 + j0.135) Ω.

6.7.2 Cables Having Uncovered Metal Sheaths

6.7.2.1 Grounding System Impedance

The calculations performed by using the previously presented analytical procedure and the given relevant data furnish the following results:

- The active cable line length is 590 m.
- The grounding impedance at the beginning of each of the connected cable lines, considered separately, is

$$Z_{c\infty} = (0.330 + \text{j}0.235)\ \Omega.$$

- The equivalent grounding impedance of the considered grounding system is

$$Z_e = (0.039 + \text{j}0.049)\ \Omega \text{ or } |Z_e| = 0.063\ \Omega.$$

The value of impedance Z_e includes the proximity effects between the substation grounding electrode and the cable lines acting as grounding conductors, as well as between the cable lines laid in the same trench.

To estimate the influence of the total proximity effect in the considered case, one can use an unrealistic assumption that the proximity effects in the considered case do not exist. Then, the grounding contribution of each outgoing cable line becomes the same, or (0.330 + j0.238) Ω, while the equivalent grounding impedance expressing grounding contributions of all of the cable lines becomes 11 (total number of cable lines) times smaller, or (0.030 + j0.022) Ω. In accordance with the presented results, the proximity effect makes the considered grounding system impedance obtain a value that is 66% greater.

To obtain a better insight into the important particularities of this type of grounding system, variations of the length L_a and impedance $|Z_{c\infty}|$ as a function of the equivalent soil resistivity for this type of cable are graphically presented in Fig. 2.17.

On the basis of the diagram in Fig. 2.17 it can be seen that the effective value of impedance $Z_{c\infty}$ is not strongly (linearly) dependent on soil resistivity as in the case of a substation grounding electrode that can be treated as equipotential on its entire surface.

6.7.2.2 Transferred Potentials

On the basis of the presented analytical expressions and the given relevant data, the transferred potential coefficient as a function of the cable line length to the nearest supplied MV/LV substation, for different soil resistivity and for different grounding impedance Z'_S, is shown in Fig. 6.17.

The curves in Fig. 6.17 are obtained under the assumption that there is only the considered cable line in the trench.

The full curves are obtained under the assumption that $Z'_S = Z_{c\infty}$ and using Eq. (6.42), whereas the broken curves result from Eq. (6.44) under the assumption $Z'_S = Z_{c\infty}/2$. On the basis of the given diagrams it is possible to note another favorable particularity of cables having uncovered metal sheaths: in conditions of low soil resistivity they significantly reduce the value of transferred potential.

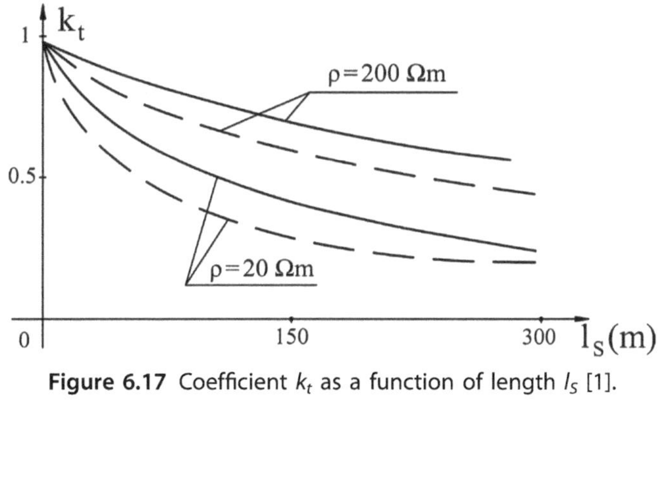

Figure 6.17 Coefficient k_t as a function of length l_S [1].

6.7.3 Cables Having Covered Metal Sheaths/Screens

According to the presented analytical procedure and the given relevant data, including those presented in Fig. 6.16, it has been obtained that the grounding contribution of the cable lines belonging to primary directions 1, 2, 3, and 4 are (0.183 + j0.139), (0.186 + j0.114), (0.182 + j0.138), and (0.287 + j0.227) Ω, respectively. Finally, the grounding contribution of all cable lines is

$$Z_e = (0.051 + j0.036)\ \Omega \ \text{ or } \ |Z_e| = 0.062\ \Omega.$$

It can be seen that the obtained value of impedance Z_e is practically the same as the value obtained in the case of cable lines with uncoated sheaths.

Variations of the length L_a and impedance $|Z_{c\infty}|$ as a function of the average grounding impedance of the connected MV/LV substations $|Z_{av}|$, whose value depends to a great extent on the degree of urbanization, are graphically presented in Fig. 6.18.

Undoubtedly, the results obtained for other practical cases would be somewhat different depending, inter alia, on the particular space dispositions of the cable lines surrounding an HV/MV substation. Namely, for the same number of outgoing cable lines the grounding contribution of the cables having uncovered sheaths are more pronounced, or the value of impedance Z_e is lower if the total length of the cables laid in the common trenches is smaller. However, in typical urban surroundings these differences cannot be so large that they have any greater influence on the main observations obtained from the presented quantitative analysis. Also since

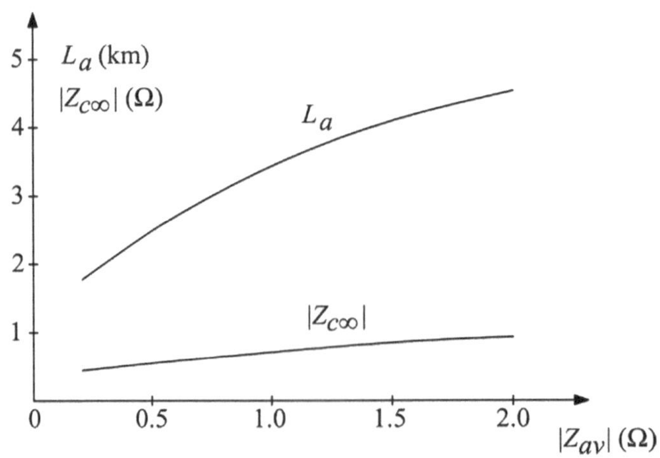

Figure 6.18 The variations of L_a and $|Z_{c\infty}|$ depending on $|Z_{av}|$ [2].

influence of the specific soil resistivity on Z_e is decreased the unavoidable arbitrariness in determining the equivalent specific soil resistivity cannot have any larger influence on the finally obtained results.

6.8 GENERAL CONDITIONS FOR SOLVING THE GROUNDING PROBLEM OF MV/LV SUBSTATIONS

The introduction of MV cables with synthetic insulation into contemporary distribution networks obviously have a certain influence on the general conditions for solving the grounding problem of MV/LV distribution substations. As first this influence is expressed through the changed value of the reduction factor of feeding lines, defined in Section 3.1. Since the value of the reduction factor, r, for the cable lines with XLPE cables is less favorable ($r \approx 0.35$ for the sheath Cu 25 mm^2 and $r \approx 0.6$ for the sheath Cu 10 mm^2) than that for paper-insulated cables ($r \approx 0.25$), it may be said that the influence of these cables to the grounding conditions of MV/LV substations is less favorable. However, this is not the only aspect of the whole problem. Grounding contributions of considered types of cables also play an important role and they should be considered as well.

The calculation of grounding effects of outgoing cable lines from MV/LV substations is basically the same as in the case of HV/MV substations. However, in this case the number of cable lines is significantly smaller and in the case of the so-called radial supply is reduced to only two lines. Then, because of the relatively small substation ground electrode, the proximity effect practically does not exist in the case of cables with uncovered metal sheaths as well. The lengths of connected cables vary from 1 to $(N - 1)$ cable line sections (Fig. 6.13). Also when MV/LV substations are located in the suburban areas, one of these two lines can be constructed as an overhead line.

When the lines are made of paper-insulated cables laid in the soil with $\rho = 50\ \Omega$m, their grounding impedance $|Z_{c\infty}|$ obtains the values 0.184 Ω for the case with two and 0.368 Ω for one cable line.

If the lines are made of XLPE cables it is obvious that the value of impedance Z_c depends on the position of the considered MV/LV substation along the feeding cable line (Fig. 6.13). Thus for the necessary calculations, Eq. (6.37) has been used.

The calculations have been based on the following relevant data:

- Cable sheath is Cu 10 mm^2 or $Z_{sh} = (0.178 + j0.211)\ \Omega$.
- The feeding cable line has 10 sections ($N = 10$).
- The average length of the cable sections is 300 m.

- The average effective value of the impedance Z_{av} is 0.5 Ω in one case and 1 Ω in the other.
- The grounding impedance of the HV substations at the cable line ends is 0.01 Ω.

The first of the values adopted for impedance Z_{av}, 0.5 Ω, corresponds to the substations located in relatively more recently built urban areas with many buildings having steel reinforcements in the concrete foundations. The greater of the adopted values of impedance Z_{av} corresponds to the substations located in the relatively older urban areas with a small number of buildings having steel reinforcement in the concrete foundations. The results of the performed calculations are shown in Figs. 6.19 and 6.20.

The used notation has the following meaning:

Z_{av}—grounding system impedance that takes into account the grounding effect of the feeding line as a connection with the grounding system of other substations, and

n_s—series number of the supplied MV/LV substation.

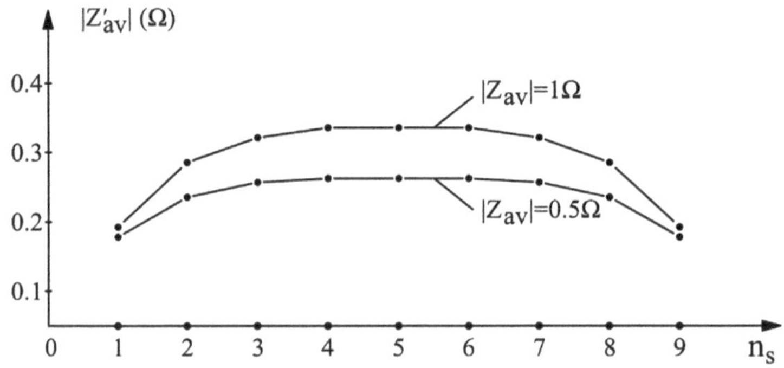

Figure 6.19 $|Z'_{st}|$ for the case of two connected MV cable lines.

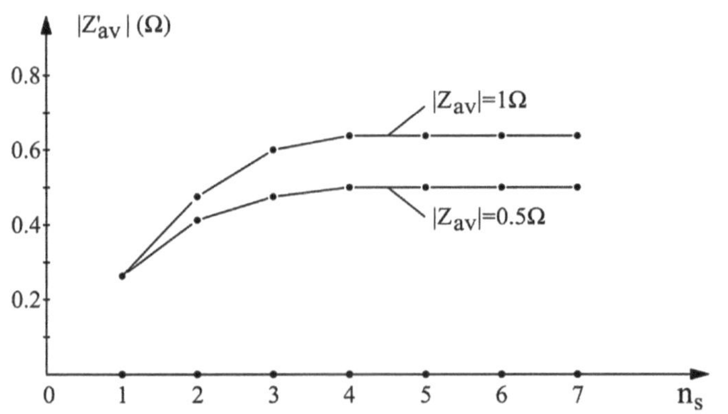

Figure 6.20 $|Z'_{st}|$ for the case of one connected MV cable line.

According to the results presented in Figs. 6.19 and 6.20 it is obvious that the grounding contributions of the connected MV cable lines are somewhat more favorable for most MV/LV substations in the case of cables with uncoated metal sheaths. However, the grounding contributions of XLPE cables are also sufficient, if one takes into account that the values of the ground fault currents in MV networks are limited to one relatively small value, usually between 300 and 1000 A.

The results of the presented analysis show the following: even XLPE cables with the smallest cross-section of metal sheath (Cu 10 mm^2) enable successful solving of the grounding problem of MV distribution substations. It is not difficult to show that by using a remarkably increased cross-section of the metal sheath (Cu 25 mm^2), grounding contributions of the connected cable lines will be only slightly improved. On the other hand, the total length of the MV cables installed each year into distribution networks is enormous. As an illustration, according to the London-based company Cables Industry Analyst CRU International, the market for power cables above 1 kV was worth US$14 billion in 2002 and a similar amount was valid for 2003. Thus any unnecessary increase in the sheath cross-section means very high additional expenses in total cable production. Besides this, it is known that in three-phase MV networks, so-called cross-bonding or metallic sheaths transposition is not applicable for the elimination of circulating currents. Therefore the unnecessary increase in the cross-section of the metallic sheath increases the losses in the electric energy caused by these currents. Since these currents exist under normal operating conditions, these losses are permanent and increase with time during the whole utilization period of a cable line. Also these unnecessary losses cause additional heating of the cables, which result in a certain decrease in the current-carrying capacity of the cable lines and whole MV distribution network. Thus it is not difficult to see that it is not rational to increase the cross-section of the metal sheath of XLPE cables to improve the general grounding conditions of the distribution HV/MV and MV/LV substations. The additional reason for this is because these cables are mainly applied in newly built urban areas, where there are many buildings with steel-reinforced foundations that improve overall grounding conditions.

The whole previous analysis is based on the assumption that in the surroundings of the considered cable lines there are no other metal installations. However, in reality, cable lines are surrounded by many metal installations typical for urban areas. Their relative participation in returning ground fault current to its sources under the same condition is certainly greater than in the case of HV cable lines (Sections 3.2 and 3.4). Thus on the basis of the presented results, one can expect elimination of the requirement for the application of expensive MV cables acting as long grounding conductors (cables with an uncovered metal sheath).

6.9 GROUNDING SYSTEM DEPENDING ON THE TYPE OF APPLIED MV CABLE LINES

A common characteristic of the considered grounding systems is that their physical dimensions are not limited to a priori and strictly defined areas. They are formed freely and spontaneously encompass various mutually connected, underground metal installations and constructions, even those that, because of the great distance from HV/MV substations, perform their grounding function at a trivial rate. Such a grounding system includes the terra-neutral in the LV network and consumer installations of many known and unknown metal installations typical for urban areas [21,22]. In this way, the area encompassed by the grounding system is always the largest possible for the certain relevant characteristics of the surrounding metal installations spontaneously serving as the elements of these grounding systems. Since the urban areas of contemporary big cities are very large, it can be said that the area involved by underground metal installations is usually significantly greater than the area where the mentioned installations perform their grounding function as the elements of the grounding system of an HV distribution substation.

The urban area involved by the considered grounding systems is larger, for certain design parameters of MV cables having an uncovered sheath, if the local soil resistivity is larger, and in cases of cables having insulated sheaths/screens, if grounding impedances of the supplied MV/LV substations are larger. This means that the considered grounding systems have the specific ability to adapt their size (dimensions) in such way that they are larger if the local soil resistivity is higher. Also, it means that these grounding systems partially compensate for the unfavorable influence of increased soil resistivity, which can appear as a consequence of a long period of dry weather.

On the basis of the performed quantitative analysis, one can see that the contribution of the HV/MV substation grounding electrode in dissipating the fault current into the earth is negligible and decreases with the increase in soil resistivity. This leads to the conclusion that it is not possible to lower significantly the value of the grounding impedance of the whole grounding system by additional interventions, e.g., by enlarging substation ground grid dimensions and by reducing its spaces, as well as by adding long vertical ground rods.

The insulated metal sheaths of a cable line act as a specific electrical line with insulated conductors. As a result, they have the lowest possible transfer impedances and a good transfer of the fault current to the distant grounding electrodes. Certainly, this intensifies the grounding function of distant underground metal installations and enlarges the area covered by the grounding system of an HV substation located in an urban area. In fact, such a grounding system represents an assembly of grounding systems composed of one main grounding electrode belonging to the considered HV/MV substation, and many small accompanied grounding systems belonging to the supplied MV/LV substations. By using Eq. (6.47) one can estimate the area covered by the considered grounding system. In the considered numerical example it is an area approximately determined by a radius of about 3490 m around the considered HV/MV substation. According to this, the covered area is approximately 29 times larger than in the same case but with cables having uncoated sheaths. Obviously, the main advantage of cables having insulated sheaths/screens emanates from the fact that the transfer impedances of their metal sheaths/screens to the supplied MV/LV substations are significantly smaller.

Some more practically possible examples will be obtained by assuming cases where the grounding conditions are identical to the considered case, but only the soil resistivity and the average grounding resistance/impedance of the supplied MV/LV substations are several times gradually increased. These cases can be considered based on the fact that long grounding conductors have a specific particularity in that they partially compensate the unfavorable influence of high soil resistivity (Figs. 2.17 and 6.17). This favorable effect in the case of the grounding systems formed with cables having insulated sheaths is more manifested because the proximity effect in the case of these cables does not exist practically. This means that for gradually increasing soil resistivity the characteristics of a grounding system formed by cables having insulated sheaths/screens will become more and more favorable in comparison to a grounding system operating under identical conditions, but utilizing cables having uncovered sheaths.

The presented quantitative analysis shows that the outgoing MV cable lines having insulated sheaths/screens participate in grounding systems of HV/MV distribution substations as good conductive connections with grounding systems of many supplied MV/LV substations. As a result, future grounding systems of HV/MV distribution substations formed only through this type of MV cable will manifest excellent grounding characteristics as well.

Under identical conditions, potentials transferred through the cables having uncovered sheaths can be significantly lower, but in many practical situations, because of the usualy large proximity of the nearest MV/LV substation, this advantage has no practical significance.

An increased use of cables having insulated sheaths/screens in MV networks will not essentially change the general conditions for solving the grounding problem of HV distribution substations. In other words, the grounding problem of these substations is not the limiting factor for a widespread application of contemporary MV cables having insulated metal sheaths/screens.

Put differently, there is no reason to fear that the wide use of cables having insulated sheaths/screens could jeopardize the possibility of achieving required safety conditions in future HV/MV distribution substations. Also, on the basis of the relatively recent investigations performed in the HV distribution network of Beograd many different metal installations in urban areas are affected by the fluctuating magnetic fields along and around the HV feeding lines. They reduce to a great extent the portion of the ground fault current dissipated into the surrounding earth through the grounding system of HV distribution substations.

REFERENCES

[1] Lj.M. Popović, Practical method for the analysis of earthling systems with long external electrodes, IEE Proceedings C 140 (1993) 213–220.

[2] Lj.M. Popović, Comparative analysis of grounding systems formed by MV cable lines with either uninsulated or insulated metal sheath(s), Electric Power System Research 81 (2) (2011) 393–399.

[3] Lj.M. Popović, Proximity effect between an earthling grid and external electrodes in an earthling system of high-voltage installations, IEE Proceedings C 133 (1986) 346–352.

[4] Lj.M. Popović, Potential distribution along external electrodes of high voltage installation grounding system, IEEE Transactions on Power Delivery 3 (4) (1988) 1508–1587.

[5] Lj.M. Popović, Active length of cable lines with uncoated metal sheaths as long grounding conductors, ETEP European Transaction on Electrical Power 22 (2) (2012) 206–215.

[6] Lj.M. Popović, Comparative analysis of grounding systems formed by feeders in one case with uninsulated and in the other case with insulated metallic sheath, in: Proc. CIRED Conference, Turin, June 2005.

[7] J.G. Sverak, Sizing of grounding conductors against fusing, IEEE Transactions on Power Apparatus and Systems 100 (1) (1981) 51–59.

[8] B. Tapar, V. Gerez, A. Balakrishman, D. Blank, Simplified equations for mush and step voltages in an ac substation, IEEE Transactions on Power Delivery 6 (2) (1991) 601–607.
[9] S.A.P. Meliopoulos, An advance computer model for grounding system analysis, IEEE Transactions on Power Delivery 8 (1) (1993) 13–23.
[10] S.T. Sobral, V.G.P. Fleury, J.R. Villalba, D. Mukhedkar, Decoupled method for studding large interconnected grounding systems using microcomputer: part I – Fundamentals, IEEE Transactions on Power Delivery 3 (1988) 1536–1544.
[11] S.T. Sobral, V.S. Costa, M.S. Campas, D. Mukhedkar, Dimensioning of nearby substations interconnected grounding system, IEEE Transactions on Power Delivery 3 (4) (1988) 1605–1614.
[12] Lj.M. Popović, Determination of the reduction factor for feeding cable lines consisting of three single-core cables, IEEE Transactions on Power Delivery 18 (3) (2003) 736–743.
[13] Lj.M. Popović, Improved analytical expressions for the determination of the reduction factor of the feeding line consisting of tree single-core cables, European Transactions on Electrical Power 18 (2) (2008) 190–203.
[14] Lj.M. Popović, Influence of the metallic installations surrounding the feeding cable line on the ground fault current distribution in supplied substations, IEEE Transactions on Power Delivery 23 (4) (2008) 2583–2590.
[15] Lj.M. Popović, Ground fault current distribution in a station supplied by a line composed of a cable an overhead section, ETEP European Transactions on Electrical Power 17 (2005) 207–208.
[16] J.E.T. Villas, D. Mukhedkar, V.R. Fernandes, A. Magalhaes, Ground grid design of a transition station system - a typical example of fault transfer, IEEE Transactions on Power Delivery 5 (1) (1989) 124–129.
[17] A. Campocia, L. Minea, G. Zizzo, A method to evaluate voltages to earth during an earth fault in an HV network in a system of connected earth electrodes of MV/LV substations, IEEE Transactions on Power Delivery 23 (4) (2008) 1763–1772.
[18] G. Zizzo, A. Campoccia, E. Riva Sanseverino, A new model for safety analysis in a system of earthing systems interconnected through bare-buried conductors, COMPEL Computation on Mathematics in Electrical Engineering 28 (2009) 412–436.
[19] IEEE Guide for Safety in Substations Grounding, Std. 80–2000.
[20] Short-Circuits Currents in Three-Phase A.C. System-Part 3: Currents During Two Separate Simultaneous Line-to-Earth Short Circuit and Partial Short-Circuit Flowing Trough Earth. IEC Int. Std. 60909–3, Ed. 3.0, 2009-03.
[21] IEC 60364 Standard: Electrical Installations Buildings.
[22] Int. Std., IEC 60364-1 Low-Voltage Electrical Installations-Part 1: Fundamental Principles, Assessment of General Characteristics, 2005.

CHAPTER 7

Voltages Induced by HV Lines Laid Through Urban Areas

7.1 INDUCTIVE INFLUENCE OF HV LINE DURING A GROUND FAULT

7.1.1 Introduction

The possibilities of appearances of dangerous and harmful voltages and currents are not limited only to areas within or in the vicinity of high-voltage (HV) substations located in urban or suburban areas. They can also appear very far from a HV substation, as a consequence of the inductive influence of its feeding line on the surrounding metal installations.

The electromagnetic interaction between an HV power line and a neighboring metal installation is divided into the following three categories: capacitive (electrostatic), conductive (resistive), and inductive (magnetic) coupling. The first is generated by an electric field, influences only metal installations above the ground, and has no path to ground. Because of the first interaction a metal installation picks up a voltage relative to the soil that is proportional to the power line voltage. The second occurs during lightning strikes or at phase-to-ground faults. When this has occurred, a large voltage cone is created around the grounding electrode as a result of the voltage surge, which can penetrate the surrounding metal installation through an installation's coating defects. The third is inductive influence, which is generated by the magnetic field present during both normal operating conditions and fault conditions when the installation is placed parallel to an HV power line. In an urban environment the inductive influence appears under very specific and complex conditions and, as a result, has only recently been solved [1−6].

When a communication line or a metal pipeline runs parallel to an HV power line over a line section, certain voltages will be induced in these installations under both normal electric power line operation and fault conditions. These voltages can be so high that they represent an electrical shock hazard for the operating personnel or general public. They are also a

Practical Methods for Analysis and Design of HV Installation Grounding Systems
ISBN 978-0-12-814460-2
https://doi.org/10.1016/B978-0-12-814460-2.00007-6

threat to the integrity of cathodic protection equipment, pipeline coating, or pipeline steel, particularly during ground fault conditions.

According to the international regulation, practice potentials in excess of 15 V should be considered hazardous and measures should be undertaken toward reducing these potentials [6]. Therefore this phenomenon deserves to be considered and clarified in all practically possible conditions and circumstances. The influence may be especially intensive in the case of an unbalanced fault in a directly earthed power network.

Prevention of both dangerous and harmful consequences of the inductive influence of HV lines on nearby insulated conductors (communication lines or different metal pipelines) has been the subject of investigations for at least the last five decades. As a result, many of the so far published papers proposed different methodologies for solving this problem. A short review of their historical development can be summarized in the following way.

The first attempts to study AC interferences were based on the widely known Carson's formulas [7]. Then, Pollaczek described in Ref. [8] inductive coupling between parallel conductors in the presence of earth as a half-space conductive medium. Paper [9] extended Carson's work to multilayer earth and conductors near point sources of current. A later paper [10] included the electrical characteristics of pipelines, such as resistance to earth of a bitumen-coated pipeline surface.

In 1986 Dawalibi and Pinho [11] described computer-based analysis aimed at ensuring personnel safety and pipeline integrity during power system faults. Then, Sobral et al. [12] applied the decoupled method to simplify the calculation of the interferences affecting a single communication circuit or a single pipeline, caused by a ground fault in a single nearby power transmission line.

Nearly 20 years ago Southey et al. [13] demonstrated, by using computer simulation software, the performance of a highly effective AC interference mitigation method. The method, which combines the effectiveness of grounding conductors and gradient control wires, attenuated both inductive and conductive interference providing, at the same time, cathodic protection. Then, a method for the calculation of currents and voltages in the system of parallel pipelines and buried conductors was proposed in Ref. [14]. The method is based on the lossy transmission line model for buried conductors and the previously mentioned methods based on the same approach, modified by abandoning the Thevenin circuit representation of pipeline sections.

The more recently published paper [15] presented a hybrid method employing finite element calculations and standard circuit analysis to calculate the induced voltages and currents in a pipeline running parallel to a faulty transmission line. Then, paper [16] developed a simulation software tool for prediction and mitigation techniques applicable to pipeline networks influenced by HV power lines. The software that can deal with any configuration (no limitation on the number of pipes, transmission lines, bonds, groundings, coating, and soil resistivity) is very user friendly and robust since a general algorithm is used to calculate the induced electromagnetic force (EMF), eliminating the need for subdivision of pipelines in sections, parallel or not, to the transmission line(s). With the calculated values of EMF, the induced voltages and currents are then obtained by solving the well-known transmission line model using a numerical technique that specifies the pipeline parameters (diameter, soil resistivity, etc.) for each individual section of the pipeline. Finally, a rigorous solution for the electromagnetic field of underground conductors and a two-layer earth is presented in Refs. [17,18].

On the basis of the previous short review, it is not difficult to observe that considerable effort has been made over many years for the development of accurate computation models for calculating voltages and currents induced by power lines in neighboring metal installations. Based on the developed mathematical models and computation techniques, several software packages are available nowadays. As an example, the method presented in Ref. [16] takes into account an arbitrary number of pipelines, power lines, bonds, groundings, coatings, and soil resistivities.

The common characteristic of current methods is that they take into consideration the constructive characteristics and mutual space disposition of all conductors involved, as well as the characteristic of the surrounding earth as a conductive medium. However, none of them produces a solution when the problem appears in urban areas, or where many relevant data are uncertain or completely unknown.

As has already been mentioned, the increasing sizes of modern power distribution networks, as well as the increasing operating and short-circuit currents of these networks, have been matched by the overspreading networks of earth return circuits (different pipelines, different cable and overhead line neutrals, etc., closely spaced to HV distribution lines). Space dispositions of all of these installations, determined mainly by dispositions of city streets, and small mutual separations result in a strong inductive and, in the vicinity of substations, conductive coupling of different network types.

The usage of common routes (mainly street sidewalks) for various supply networks (electricity, water, gas, oil, telecommunications, etc.) unavoidably leads to the appearance of their mutually inductive interaction. From the standpoint of current engineering practice the whole problem has three different aspects. Considered separately, they can be defined in the following manner:

- Influence of the surrounding metal installations on the fraction of ground fault current dissipated into the surrounding earth through the grounding system of a supplied substation, or grounding problem of supplied HV/MV substations located in urban or/and suburban areas,
- Influence of the surrounding metal installations on the transfer characteristics of power lines passing through urban and suburban areas, and
- Impact of the surrounding metal installations on the magnitude of the inductive influence of an HV line on any of these installations seen separately.

The problem of determining the influence of metal installations surrounding an HV feeding cable line on ground fault current fraction dissipated through the grounding system of a supplied distribution substation is considered in Ref. [1]. The solution is obtained by introducing a fictitious conductor of cylindrical form placed around and along the entire feeding cable line. Somewhat later, this solution has been extended to include the HV overhead lines as feeding distribution lines [2]. Then, the method for determining the actual value of the ground fault current portion, which in urban conditions circulates only through the earth, has been somewhat simplified by introducing an equivalent conductor substituting all surrounding metal installations [3].

The investigation results presented in Refs. [1,2] show that a portion of the ground fault current flowing through the earth in typically urban environments is 3–5 times smaller than it was considered earlier. Also the solution achieved in Refs. [1,2] simultaneously gives a possibility for solving another practical problem that has also not been solved in the past. This is the problem of determining the series impedance of distribution lines without ignoring the presence of the surrounding metal installations. These installations act as additional neutral conductors of any distribution line passing through urban and/or suburban areas and, in accordance with this, improve its transfer characteristic. This very useful effect depends on the types of distribution lines (overhead or underground) and on the degree of urbanization of the area through which the lines pass [4].

The solution to the problem of inductive influence is obtained in Ref. [5] by further developing the method presented in Refs. [1,2]. This has been done by introducing into the calculation procedure an equivalent conductor that can be, for the considered induced installation, treated as an equivalent screening conductor. It substitutes, from the standpoint of inductive influence, all surrounding metal installations, including the inducing line neutral conductor(s). By introducing this equivalent conductor, it is possible to determine the current induced in any of the surrounding metal installations separately, provided that its technical characteristics and space disposition with respect to the considered inducing (power) line are known [5]. However, the screening factor (e.g., Ref. [19]) determined on the basis of this fictitious conductor represents an average screening factor applicable as the exact one only in the case where an induced installation is exposed along the whole length of the inducing line. In other words, its value is not quite correct in cases where the induced installation is exposed only along one section of the inducing line.

Certainly, without the possibility of correctly predicting the local screening factor [6,19], one is left without correct values of the voltages and currents that can appear in a planned metal installation that will be under inductive influence. Since the ground fault current in an HV distribution network can be in excess of 20 kA, this means that dangerous potential differences and harmful induced voltages can appear at different places without our knowledge and/or control. In other words, the necessary protective measures can be omitted in cases where they are really indispensable. Also, as a consequence of an incorrect calculation procedure, protective measures can be applied in cases where they are not necessary.

The solution to this problem is presented in Ref. [20]. It is based on developing the analytical expression for the resulting screening factor that involves inductive influences of all surrounding metal installations present only within the section of exposure. The equivalent screening conductor is imagined as a cylinder situated around of the planned/designed installation along the section of exposure.

The developed method is based on the analytical expressions for the self- and mutual impedances of line conductors installed in the semiinfinite and homogeneous earth, e.g., Ref. [2]. It enables determination of the voltages and currents induced in any metal installation planned in a concrete urban environment and in parallel with an HV line.

The presented method also solves any problem of inductive coupling that can be encountered anywhere in practice, without a need for collecting numerous input data and by using a relatively simple set of measurements as well as a relatively simple calculation procedure.

7.1.2 Description of the Basic Problem

In rural areas due to strict environmental regulations, HV power lines and other types of metal installations are forced to share a common corridor called a right-of-way and because of this these installations are very often mutually close. Also in the case of urban and suburban areas such situations are practically unavoidable and HV power lines share the same streets with many different urban installations. As a final consequence of this, too high voltages can be induced in each metal installation located along and nearby an HV line, particularly in cases such as long exposition or an unbalanced fault in the power system.

As we have seen in Chapter 3, during a ground fault in a directly and indirectly earthed HV network, the fault current is returning toward its sources by using all available conductive paths. These paths are provided by shield wire(s) and a counterpoise conductor in the case overhead lines and by cable sheaths, metal armor, and pipes in the case of underground lines. Parts of ground fault current also return through different surrounding metal installations that are not foreseen for such function. In present-day urban areas these installations can be gas and/or water pipelines, other electric power line neutral conductors, and metal sheaths of telecommunication cables.

Under normal operating conditions an unbalanced current also circulates through the earth and neutral conductor(s), including the surrounding metal installations. Its value is certainly significantly smaller in comparison with the current appearing during a ground fault, but it flows permanently and because of this it can also be harmful and disturbing in some cases. As is well known, more sensitive electronic systems face an increasingly hostile electromagnetic environment. Magnetic coupling is an increasingly frequent cause of problems capable of upsetting data transmission. Thus close attention is required for determination of inductances of the circuit loops considered.

By disregarding for a while all other surrounding metal installations normally existing in urban areas, let us assume an HV line connecting substations A and B and having one section running along the same street(s) as a planned metal installation. Such a situation is schematically depicted in Fig. 7.1.

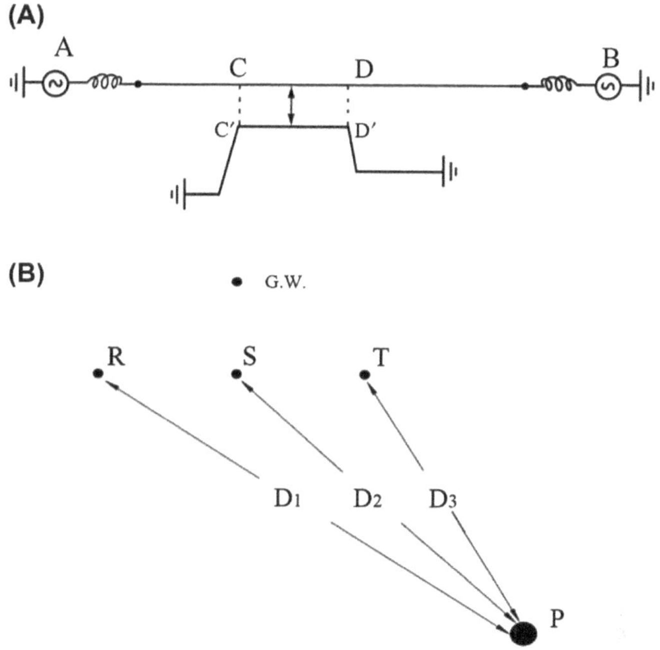

Figure 7.1 Typical situation of inductive coupling [20]. (A) Section of exposure to an inductive influence. (B) Separations of overhead line conductors and the exposed installation.

The used notation has the following meaning:

C and *D*—ends of the inducing line section that correspond to the ends of the exposed section of the induced installation,

C′ and *D′*—ends of the exposed section of the induced installation,

R, S, and *T*—phase conductors,

P—pipeline or some other type of metal installation,

D_1, D_2, and D_3—separations between the induced installation and the individual phase conductors, and

G.W.—ground wire of an HV overhead line.

The considered induced installation is insulated from earth all along its length and earthed only at its ends. Grounding systems of a distribution HV/MV substation located in urban areas consist of the substation grounding electrode and many outgoing MV cable lines acting as external grounding electrodes and/or as conductive connections to the grounding systems of the supplied customer substations. Thus a spontaneously formed grounding system involves a large urban area around an HV/MV substation. This grounding system includes the terra-neutral earthing system of the low-voltage (LV) networks and consumer installations and many, known and unknown, metal installations typical for urban surroundings. As a consequence, the outgoing MV cable lines simultaneously become the conductive connections with the metal installations laid along the same street(s) as the HV feeding line. Thus it is not difficult to see that each unbalanced current in an HV feeding line for its return into the power system uses earth as a conductive medium, line neutral conductor(s), and all other available paths, the metal installations surrounding this line (Fig. 3.6). Thus it can be said that this current is only partially circulated through the grounding system of the supplied substation and earth, while the rest of this current returns to its sources in the power system through the line neutral conductor(s) and all metal installations surrounding the considered HV feeding line. Since the effective load currents of HV distribution lines are practically symmetrical under normal operation conditions, an unbalanced current appears only as a consequence of different space positions of the phase conductor in relation to the neutral conductor(s) and surrounding metal installations.

As an illustration, the main current fractions appearing under ground fault conditions within and around an HV distribution line are presented in Fig. 3.6.

On the basis of the given description, it is not difficult to see that each of the surrounding metal installations acts as a screening conductor in relation to the current and voltage induced in any of these installations seen individually. Thus the considered problem can be defined in the following, more explicit, way: how can the screening effect of all, known and unknown, surrounding metal installations be determined?

The first practical difficulty stems from the fact that one of the basic parameters necessary for estimation of these mutual interactions, soil resistivity of the surrounding area/earth, cannot be determined. Although there are several methods of measuring soil resistivity, none of them is applicable to urban conditions. This is because surfaces of urban areas are covered with buildings, streets, sidewalks, and many other permanently constructed objects. Also belowground, many known and unknown metal installations have already been laid. However, the main practical difficulty is caused by the fact that the surrounding metal installations are mostly laid under the street sidewalks and many relevant data about them cannot be visually determined or verified. As a result, these data remain hidden and partly or completely unknown for the designers of a planned metal installation. The only known fact is that some of the surrounding metal installations are not in direct contact with the earth, while others are in direct and continuous contact with the earth. Therefore it is also known that the spatial dispositions and structures of these installations are dependent on many local factors and, as a result, they are different from case to case. In addition, many of these installations are purposely/intentionally conductively connected. Therefore they spontaneously form a very large and complex underground metal network whose structure and topology vary along each of the HV distribution lines. As a result, the screening effect of the surrounding metal installations is different for different sections exposed to the same HV line. Thus this desirable effect should be determined separately for each concrete section of exposure.

Each of the surrounding neutral conductors or metal installations, together with earth as the common return path, forms one electrical circuit, while all of them together (line neutral conductor[s] and all surrounding metal installations) form inductively coupled electrical circuits, whose number is equal to the number of these conductors and installations. Since these circuits can be represented by the corresponding system of equations and since the analytical expressions for the self- and mutual impedances of these conductors and metal installations are known, it can be said that the considered problem was in principle solvable a long time ago. However, this solution is not sufficient for all practical problems because of uncertain and unknown relevant data, including in some cases the number of the surrounding metal installations involved. Thus because of practical difficulties the considered problem can be summarized in the following question: how can the objective deficiency of numerous relevant data be compensated for?

7.2 INDUCTIVE INFLUENCE ALONG THE WHOLE LENGTH OF AN HV LINE

The longitudinal EMF induced along any of the surrounding metal installations cause certain current circulations and a potential between this metal installation and the earth immediately outside the coating. As we have already seen in Section 3.2, currents induced during a ground fault in the metal installations surrounding an HV feeding line substantially reduce all potentials and potential differences within the grounding system of an

HV substation located in an urban or suburban environment. However, this very favorable effect is accompanied by one that cannot be considered as desirable. Namely, we have to know whether the potentials that appear along any of the surrounding metal installations are hazardous if the grounding electrode is not applied at places where these installations can be exposed to human touch. As is well known, such a situation is dangerous mainly for those working on the maintenance of these installations.

The formerly described problem in urban areas is not so pronounced because all surrounding metal installations exhibit the very favorable, so-called screening, effect and can considerably reduce the induced EMF. This effect is expressed through the screening factor of electric power line neutral conductors. In the case of any electrical line having neutral conductor(s) this coefficient is defined as the ratio of the EMF induced by a line equipped with neutral conductor(s) and the EMF induced by this line, but without neutral conductor(s) [19].

To obtain a better insight into this effect in cases where the inducing line is surrounded by many metal installations acting as additional line neutral conductors, the completed equivalent circuit and an isolated metal installation planned to be laid along the whole HV line length is presented in Fig. 7.2.

The used notation has the following meaning:

$V_{(N+1)}$—potential that will be induced along the whole length of a planned installation,

U_{nT}, U_{n1}, U_{n2}, …, U_{nN}—voltages induced in an arbitrary (*n*th) metal installation by the currents in the surrounding circuits,

I_F—ground fault current through the nearest (critical) phase conductor T,

I_1, I_2, I_3, …, I_n, …, I_N—currents through the neutral conductor(s) and surrounding metal installations,

Z_1, Z_2, Z_3, …, Z_N—self-impedances of the line neutral conductor(s) and individual surrounding metal installations, and

N—unknown total number of surrounding metal installations enlarged by the number of line neutral conductors.

The method for determining voltages induced during a ground fault by an HV line passing through an urban area in any of the surrounding metal installations following this line along the whole of its length is presented in Section 3.2. By this method, all of the line neutral conductors and all surrounding metal installations, excluding the planned one, are substituted by only one equivalent line neutral conductor. The adopted space

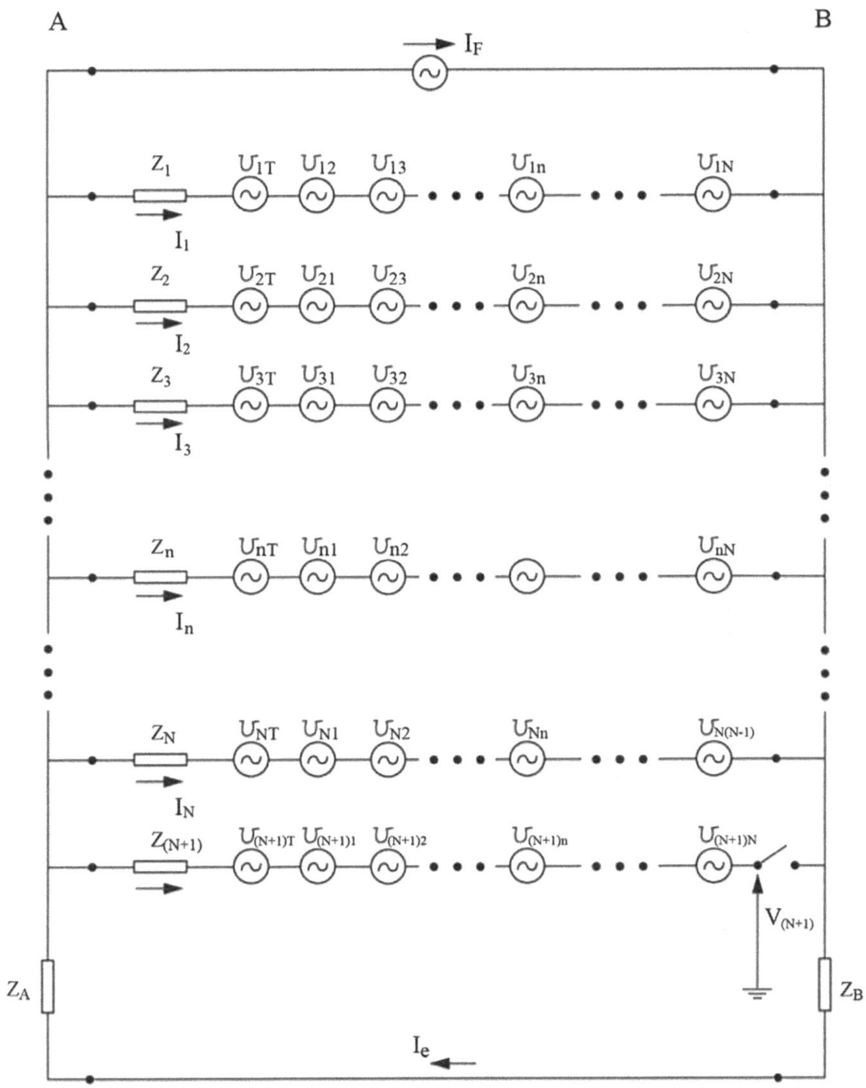

Figure 7.2 Complete equivalent circuit during a ground fault. *(Reproduced courtesy The Electromagnetic Academy: Lj.M. Popović, Ground fault current distribution when a ground fault occurs in HV substations located in an urban areas, Progress In Electromagnetic Research B 59 (2014) 167–179.)*

disposition with respect to the line phase conductors and physical appearance of this imagined conductor are such that analytical expression for the actual reduction factor of the power line simultaneously represents the screening factor of the equivalent conductor. Thus the resulting screening factor of certain line neutral conductor(s) and all surrounding metal installations in this case of exposure are, according to expression (3.28), determined by:

$$k = r_a = 1 - \frac{R'_C Z_{T1} + (Z_1 - Z_{1C}) Z_{TC}}{Z_1 Z_C - Z_{1C}^2}, \tag{7.1}$$

where Z_{T1} is the mutual impedance between the phase conductor carrying the simulated ground fault current and the neutral conductor used for on-site measurements, Z_{TC} is the mutual impedance between the phase conductor carrying the simulated ground fault current and the introduced equivalent conductor determined by using the procedure presented in Section 3.2 (Ω), Z_1 is the self-impedance of the neutral conductor used for on-site measurements (Ω), Z_C is the self-impedance of the introduced equivalent neutral conductor substituting all surrounding metal installations (Ω), and Z_{1C} is the mutual impedance between the neutral conductor used for on-site measurements and the introduced equivalent neutral conductor substituting all surrounding metal installations (Ω).

When the resulting screening factor k is determined, the potential that will be induced at any point in the designed metal installation can be determined by the following simple expression:

$$V_i(l) = klZ'_{T(N+1)} I_F, \tag{7.2}$$

where I_F is the maximal anticipated ground fault current through the nearest (critical) phase conductor T (A), usually obtained by a previously performed separate study, $Z'_{(N+1)T}$ is the mutual impedance between the phase conductor T and the planned metal installation (Ω/km), and l is the length of the planned metal installation from substation A to an arbitrary point along this installation (km).

Thus it becomes clear that if current I_F and the space position of the designed installation are known, to determine the induced voltage at any point along the planned installation it is sufficient to determine, in accordance with the procedure presented in Section 3.2, the value of the actual reduction factor of the considered HV line.

Under normal operating conditions the phase conductors of the power line carry currents I_R, I_S, and I_T. These currents are significantly lower compared to the currents at fault conditions and their influences on the nearby installations cancel each other to a great extent. Thus the whole inductive influence during normal operating conditions is the result of the differences in the separations of the different phase conductors from the considered surrounding metal installation (Fig. 7.1B). This type of problem certainly can be solved by the procedure analogous to the procedure presented here [20].

7.3 INDUCTIVE INFLUENCE ALONG ANY SECTION OF AN HV LINE

Data regarding the magnitude of the actual inductive influence of an HV line are necessary at the design stage of a future nearby installation. For a certain value of the ground fault current in the inducing line this influence is dependent on the number and spatial positions of the line neutral conductors, all local surrounding metal installations, and local soil resistivity. However, in practical conditions it is not possible to collect all necessary data. As a result, the possibility of solving the problem is seen in the fact that, according to the equivalent circuit given in Fig. 7.2, the total voltage induced in each of the surrounding installations involves cumulative inductive influence of all other known and unknown surrounding metal installations. This indicates that the problem of deficiency of many relevant data can be overcome only by performing the corresponding on-site measurements (Section 3.2). However, at the design stage the planned installation does not exist. Thus for the purpose of performing the necessary measurements one can use one provisory single-core cable line instead of the planned installation. It can be performed by laying a cable line on the surface of the soil exactly along the foreseen section of the parallelism, from C' to D' (Fig. 7.1). For this purpose it is best to use an LV single-core cable (with no metal sheath), suitable for application in different urban conditions and circumstances. This provisory cable line should be earthed only at one end with the aim of forming, together with the earth as the return path, the open electrical circuit schematically presented in Fig. 7.3.

Figure 7.3 Provisory electrical line necessary for performing on-site measurements [20].

The used notation has the following meaning:
U_{im}—total EMF induced in the provisory electrical circuit (V), and
V—voltmeter.

From the equivalent circuit of the HV distribution line and the surrounding metal installations presented in Fig. 7.2, total voltage induced in the provisory cable line, denoted as $(N+1)$, is:

$$U_{im}(I_f) = L_{CD} Z'_{(N+1)T} I_f + L_{CD} \sum_{n=1}^{n=N} Z'_{(N+1)n} I_n(I_f) \tag{7.3}$$

where I_f is the measured test current obtained through phase conductor T by simulating a ground fault in supplied substation B, whereas L_{CD} denotes the exposure length.

Since impedance $Z_{(N+1)T}$ is known when the relevant data regarding the space position of the designed installation are known (D_3 in Fig. 7.1) and since quantities I_f and U_{im} can be obtained by measurements, all these quantities can be considered as a priori known. Then, according to Eq. (7.3), total voltage induced due to the simulated ground fault is:

$$\sum_{n=1}^{n=N} Z_{(N+1)n} I_n(I_f) = U_{im} - Z_{(N+1)T} I_f. \tag{7.4}$$

The magnitude of the screening effects is expressed as the amount of reduction of the value of the electrical quantity, which is considered representative of the induction in the inducted line due to the screening current(s) [19]. Thus, according to Eq. (7.3), the resulting screening factor is determined by:

$$k(I_f) = 1 + \frac{\sum_{n=1}^{n=N} Z_{(N+1)n} I_n(I_f)}{Z_{(N+1)T} I_f}. \tag{7.5}$$

On the basis of Eq. (7.4), expression (7.5) takes the following form:

$$k(I_f) = \frac{U_{im}(I_f)}{Z_{(N+1)r} I_f}. \tag{7.6}$$

Metal installations in urban areas are mainly under the soil surface and as a consequence the resulting screening factor obtained by using Eq. (7.6) is slightly smaller than the actual one in the case of an HV overhead line and slightly greater than the actual one in the case of an HV cable line.

When the value of the resulting screening factor is determined by on-site measurement the maximal voltage that will be induced in the designed metal installation during a ground fault is:

$$U_i(I_F) = k(I_f)Z_{(N+1)T}I_F. \tag{7.7}$$

In accordance with the given expression, once the value of the resulting local screening factor is determined, to determine the voltages that will be induced in the designed installation under ground fault conditions it is necessary to calculate only mutual impedance between this installation and the nearest phase conductor of the inducing line.

It is common practice to consider the induced electromotive force as the most important quantity caused by induction. When it is determined and when the self-impedance of the designed metal installation and grounding impedances/resistances at its ends are also determined, the current that will be induced in the planned installation can also be predicted.

In this way the influence of the earth as a conductive media and the influences of all surrounding metal installations due to the fluctuating magnetic field around and along the HV power line are cumulatively involved through the local resulting screening factor, k. In other words, when this factor is determined according to the presented procedure, calculation/determination of the inductive influence in each concrete case is reduced to the simplest case or to the case where only one inducing and one screening conductor exist. However, in cases where the predicted inductive influence is considered too high and the protective measures are mandatory, it is necessary to determine the relevant parameters of this equivalent screening conductor.

7.4 DETERMINATION OF EQUIVALENT SCREENING CONDUCTOR

If the predicted values of the voltage or current are in a certain case too high, the measurements for their mitigation can be indispensable. Such measurements usually represent a purposefully laid, so-called, compensating conductor [19]. However, in an urban environment the actual effects of these measurements cannot be determined by the so far known and used method. As well as in the former case the problem was unsolvable because of many uncertain and unknown but relevant data.

With the aim of simplifying the problem, let us imagine an equivalent conductor substituting, from the standpoint of the screening effect, all

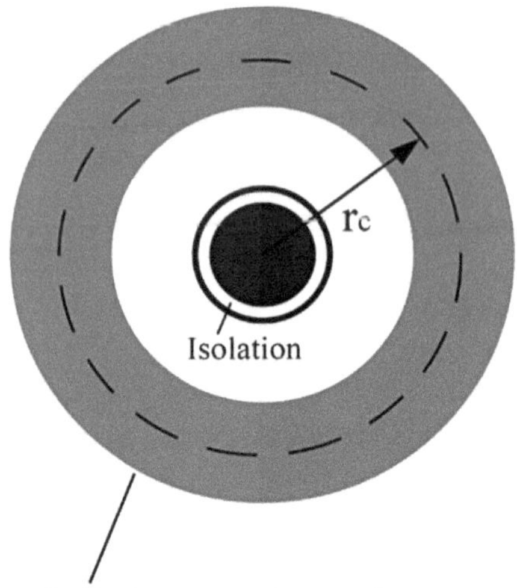

Figure 7.4 Cross-section of the planned installation and its equivalent screening conductor [20].

actually existing surrounding metal installations. The problem of determining such a conductor becomes solvable when it is assumed that this equivalent conductor is of cylindrical form and placed around the inducing line along its entire length [1–5]. However, in this case we will assume that such conductor is placed around the designed metal installation and only along its section of exposure. Also it is assumed that this conductor is perfectly grounded at both ends. Under these assumptions the planned metal installation and its imagined screening conductor are schematically presented by the model shown in Fig. 7.4.

On the basis of the presented physical appearance of the imagined equivalent conductor, the corresponding equivalent circuit of one inducing conductor (e.g., phase conductor T with current I_f) and the formerly described provisory cable line becomes essentially simplified, as shown in Fig. 7.5.

The used notation has the following meaning:

$U_{(N+1)T}$—voltage induced by current I_f in the provisory cable line,

$U_{(N+1)C}$—voltage induced by current I_C in the provisory cable line,

U_{CT}—voltage induced by current I_f in the equivalent screening conductor,

I_f—simulated ground fault current,

I_C—current passing through the equivalent screening conductor,

Z_C—self-impedance of the equivalent screening conductor,

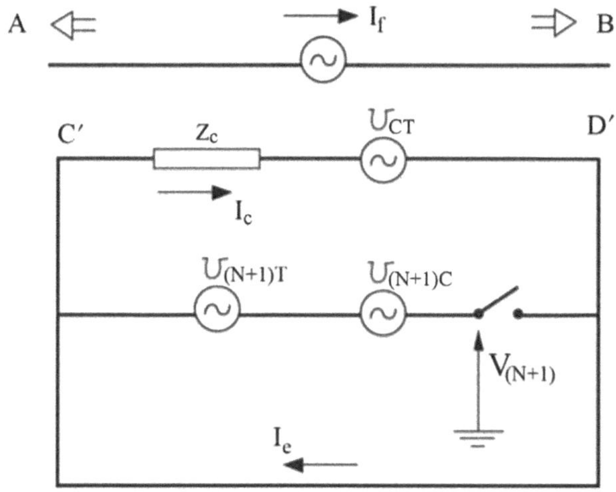

Figure 7.5 The simplified equivalent circuit of the provisory cable line [20].

Z_{CT}—mutual impedance between the equivalent screening conductor and the phase conductor T (Ω), and

U_{im}—measured value of the voltage induced in the provisory cable line (Fig. 7.5) (V).

The relevant parameters of the introduced equivalent conductor can be determined under the assumption that the voltage induced in the provisory cable line remains unchanged by introducing this conductor. On the basis of the equivalent circuit presented in Fig. 7.5, this condition can be expressed by the following system of equations:

$$\begin{aligned} &1.\ Z_{(N+1)T} I_f + Z_C I_C = 0 \\ &2.\ Z_{(N+1)T} I_f + Z_{(N+1)C} I_C = U_{im} \end{aligned} \tag{7.8}$$

where Z_C is the self-impedance of the equivalent conductor (Ω).

For the adopted physical appearance of the equivalent conductor, impedances Z_C, $Z_{(N+1)T}$, and Z_{CT} are a priori known and determined by:

$$Z'_C = R'_C + \omega\frac{\mu_0}{8} + j\omega\frac{\mu_0}{2\pi}\ln\frac{\delta}{r_C},\ \Omega/\text{km}, \tag{7.9}$$

$$Z'_{(N+1)C} = Z'_{CT} = \omega\frac{\mu_0}{8} + j\omega\frac{\mu_0}{2\pi}\ln\frac{\delta}{r_C},\ \Omega/\text{km}, \tag{7.10}$$

where R'_C is the resistance of the equivalent screening conductor (Ω/km) and r_c is the medium radius of the cylinder representing the equivalent screening conductor (m).

Since voltage U_{im} is measured and impedances $Z_{(N+1)T}$ and Z_{CT} can be determined by calculation, these quantities can be considered as known and the formerly defined condition Eq. (7.8) can be expressed in the following, simpler form:

$$\frac{Z_C}{Z_{(N+1)C}} = \frac{U_i}{U_i - U_{im}}, \tag{7.11}$$

where U_i is the voltage that would be induced assuming that the screening conductors do not exist (V).

Since expression (7.11) gives a relationship between complex quantities, it can be represented by the following system of two equations:

$$\begin{aligned} &1.\ \mathrm{Re}\{(U_i - U_{im})Z'_C\} = \mathrm{Re}\left\{U_i Z'_{(N+1)C}\right\}, \\ &2.\ \mathrm{Im}\{(U_i - U_{im})Z'_C\} = \mathrm{Im}\left\{U_i Z'_{(N+1)C}\right\}. \end{aligned} \tag{7.12}$$

Then, the relevant parameters of the introduced equivalent screening conductor (R'_C and r_c), on the basis of Eqs. (7.5–7.7), can be determined by:

$$R'_C = \mathrm{Re}\left\{\frac{U_i}{U_i - U_{im}} Z'_{(N+1)C}\right\} - \omega\frac{\mu_0}{8}, \tag{7.13}$$

$$X'_C = \mathrm{Im}\left\{\frac{U_i}{U_i - U_{im}} Z'_{(N+1)C}\right\}, \tag{7.14}$$

$$r_c = \frac{\delta}{e^{2\pi X'_C/\omega\mu_0}} \tag{7.15}$$

where $R'_C(X'_C)$ is the real (imaginary) part of the self-impedance of the equivalent conductor (Ω/km).

After determining the relevant parameters of the equivalent screening conductor (R'_C and r_c), Eqs. (7.9) and (7.10) can be used to obtain Z_C and $Z_{(N+1)C}$, as well as I_c and I_e (Fig. 7.5). Since voltages U_i and U_{im} are the phasor quantities, to obtain more accurate results it is also necessary to determine the phase angle between them. This can be done by measuring the phase angle between U_{im} and I_f and by the corresponding calculation based on the equivalent circuit of Fig. 7.5.

When the relevant parameters of the equivalent screening conductor are known, by using an analogous calculation procedure and a prior determined space position of the compensating conductor (usually just behind the planned installation), one can determine the maximal permissible value of the total impedance of the circuit formed by the compensation conductor. On this basis one can adopt the corresponding conductor that should be installed to reduce the induced voltage within the prescribed limits and ensure the necessary safety conditions and/or equipment integrity. By taking into account the screening effects of all known and unknown surrounding metal installations, this conductor can be selected in accordance with actual needs.

On the basis of expressions (7.10), (7.13), and (7.15), the actual value of the equivalent soil resistivity is necessary for determining $Z_{(N+1)R}$, $Z_{(N+1)S}$, $Z_{(N+1)T}$, R_C, and r_c. However, although there are several methods of measurement of soil resistivity (e.g., Ref. [21]), none of them is applicable in urban conditions. This is because the surfaces of urban areas are occupied by buildings, streets, sidewalks, and many other permanently constructed objects, while belowground many known and unknown metal installations are already present and unavoidably influence the results of measurements. Thus we are forced to adopt an approximate value of the equivalent soil resistivity, based on the main geological characteristics of the surrounding area, and use it in the necessary calculations. Here the favorable circumstance is that the mentioned quantities are, according to the given analytical expressions, only slightly dependent on the equivalent soil resistivity and more accurate data regarding this parameter do not have any larger practical significance. It is sufficient to know that, within the range of possible values, the lowest one should be preferred because it gives a final result that is slightly on the safe side. Thus it can be said that accuracy of the developed method depends mainly on accuracy of the measurement of the voltage U_{im} and phase angle between this voltage and the test current in the inducing line.

Since the presented method is based on on-site measurement data, it takes into account all relevant factors and parameters, including those whose influence is very small. Some inaccuracy can appear only as a consequence of the inductive influence of the nearby MV and LV distribution lines. If necessary, this influence can be efficiently avoided by using a test current of somewhat higher frequency, which can easily be discriminated from the omnipresent power frequency. The error introduced in such a way is small and its influence on the final calculation results are such that they are slightly on the safe side.

7.5 QUANTITATIVE ANALYSIS

7.5.1 Method Verification

For this purpose let us assume one 110 kV distribution line consisting of three single-core cables laid in a triangular formation along the whole line length. The phase conductors of these cables are made of aluminum having a cross-section of 1000 mm^2, while their metal sheaths are made of copper strings having a total cross-section of 95 mm^2, and the medium diameter is equal to 91 mm. Also we will assume that the equivalent soil resistivity of the area through which this line is laid is 30 Ωm.

Now, to verify the developed method, let us assume that a ground fault current of 20 kA circulates through one of the line cables, while the metal sheath of one of the other two cables is not grounded at one of the line ends. Also it will be assumed that there are no other metal installations in the surrounding space. In accordance with the previous assumptions, this HV line is presented by the equivalent circuit shown in Fig. 7.6.

By using the existing method based on the known relevant data, the given equivalent circuit, and the corresponding closed system of equations (e.g., Ref. [6]), it can be shown that the voltages induced in the metal sheath having one end not grounded take the following values: $U_i = (986 + j10846)$ V/km and $U_{im} = (1735 + j276)$ V/km. Then, by using these results as input data for the method presented here, it can be shown that the relevant parameters of the equivalent screening conductor substituting two grounded metal sheaths are $R'_C = 0.0936$ Ω/km, $r_c = 66.1$ mm. Then, on the basis of these data, it is not difficult to show that the resulting screening factor of the two cable sheaths in the considered case is:

$$k = (0.0438 + j0.1678).$$

Finally, by using this value the same value of the induced voltage has been obtained as the one obtained by using the existing method and the

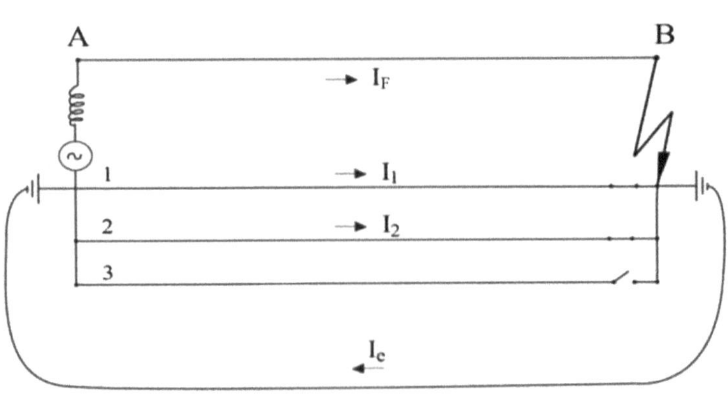

Figure 7.6 The equivalent circuit of the adopted line [20].

known data regarding the considered line. Thus it can be said that the accuracy of the presented method has been verified.

7.5.2 Screening Effects of the Surrounding Metal Installations

With the aim of obtaining an engineer's feeling about the magnitude of the possible screening effects of the surrounding metal installations in the formerly considered cable line case, let us assume that the considered line is laid in an urban environment. The desired insight will be obtained on the basis of the results of the experimental investigations presented in Section 3.4. These investigations performed on two 110 kV distribution lines of different lengths show that the relevant parameters of the equivalent conductor substituting all surrounding metal installations are the following:

- In the case of the line passing through the lower degree of urbanization: $R'_C = 0.08\ \Omega/\text{km}$, $r_c = 12.3$ m, and
- In the case of the line passing through the higher degree of urbanization: $R'_C = 0.053\ \Omega/\text{km}$, $r_c = 8.1$ m.

On the basis of these data, it is not difficult to show that the considered induced voltage is smaller by 22.8% in the case of the lower and 47% in the case of the higher degree of urbanization. This certainly leads to the general conclusion that the inductive influence of HV distribution lines depends of the degree of urbanization and that it is lower where the degree of urbanization is higher.

Also on the basis on the given data it is not difficult to determine the maximum stress voltages to which the cable sheath coating will be exposed at the critical place, the middle of the line, since sheaths are earthed at both line ends. For the shorter line (length: 2320 m) they are:

- In the case without surrounding metal installations: (2012.6 + j320.2) V,
- In the case of the lower degree of urbanization: (837.7 + j1319.7) V, and
- In the case of the higher degree of urbanization: (1072.8 + j120.4) V.

For the longer line (length: 6590 m) they are:

- In the case without surrounding metal installations: (5716.8 + j909.4) V,
- In the case of the lower degree of urbanization: (2379.4 + j3748.7) V, and
- In the case of the higher degree of urbanization: (3947.2 + j342.0) V.

Then, it will be assumed that the ground fault current passing through the considered lines is 20 kA and that the planned metal pipeline will be laid in the same sidewalk at a distance of 1 m from the nearest phase conductor.

On the basis of Eq. (7.2) and the reduction/screening factor determined by the analytical expression given in Section 3.1, one obtains that the voltage induced in the designed pipeline will be 822.5 V/km. However, if one takes into account the screening effects of the surrounding metal installations, this value is significantly smaller, i.e., 282 V/km for the shorter and 227.2 V/km for the longer line. If one assumes that the separation between HV line and the planned pipeline is 20 m, the corresponding induced voltages under all other identical conditions would be 731 V/km, 250.6 V/km, and 201.9 V/km, respectively.

The presented results show that the screening effect of the surrounding metal installations is significant and, as a result, should be taken into consideration when one evaluates the voltages induced by ground fault current passing through HV power lines in urban areas.

7.6 GENERAL OBSERVATIONS

The presented method enables the determination of screening effects of many different, known and unknown, surrounding metal installations, reducing the magnitude of the inductive influence of HV power lines passing through urban areas. In each concrete case these installations can be completely represented by only one equivalent screening conductor. This is enabled by the here described on-site measurement procedure and by the corresponding analytical method based on the results of these measurements. By determining the relevant parameters of this conductor, each practical situation, irrespective of its complexity and the amount of uncertain and completely unknown but relevant data, can be transformed into the simplest case from the standpoint of necessary calculations. From a mathematical point of view, it is interesting to note that this method offers a solution in cases where, because of the lack of a large amount of relevant data, the necessary system of equations enabling application of the previously known method cannot be completed.

On the basis of the determined equivalent screening conductor relevant parameters, it is not difficult to see that the potentially dangerous and harmful influence of an HV line upon any newly built or existing surrounding installation considered individually is smaller if the degree of urbanization is greater. In other words, when the number of electromagnetically involved metal installations is greater, this influence upon each of them, seen separately, is smaller.

In unsettled areas due to strict environmental regulations, HV power lines and other types of metal installations are forced to share a common corridor called the right-of-way and as a result these installations are very often mutually close. As a consequence of this, too high voltages can be induced in each metal installation located along and close to an HV line, particularly in cases such as a long length of exposure and/or a high level of anticipated ground fault currents. In such environmental conditions the presented method also enables solving the problem without the need to collect numerous relevant data and the application of a complicated mathematical model based on the possibilities of modern computers. All necessary calculations can be implemented by using handheld calculators only.

The presented method enables solution to the problem of predicting induced voltages due to electromagnetic fields of power lines passing through any (urban or rural) areas by compensating for the deficiency of an unlimited amount of relevant data. The final results are obtained by performing relatively simple on-site measurements and applying a relatively simple calculation procedure. Bearing in mind all these facts the presented method could be used as a foundation for the revision of current versions of the corresponding technical standards [19,22,23].

With a full knowledge of electromagnetic induction, utilities can be allocated acceptable rights-of-way for their power lines, industrial and commercial buildings can be allocated sufficient distances from power lines, housing projects can determine the heights of their buildings, and utilities can be given good separation of their pipelines and railroads from the power lines involved.

REFERENCES

[1] Lj.M. Popović, Influence of metal installations surrounding the feeding cable line on the ground fault current distribution in supplied substations, IEEE Transactions on Power Delivery 23 (4) (2008) 2583–2590.

[2] Lj.M. Popović, Testing and evaluating grounding systems for substations located in urban areas, IET Generation Transmission and Distribution 5 (2) (2011) 231–238.

[3] Lj.M. Popović, Determination of the actual reduction factor of HV and MV cable line passing through urban and suburban area, FACTA UNIVERSITATIS-Series: Electronics and Energetics 27 (1) (2014) 25–39.

[4] Lj.M. Popović, Transfer characteristics of electric-power lines passing through urban and suburban areas, International Journal of Electrical Power and Energy Systems 56 (0) (2014) 151–158.

[5] Lj.M. Popović, Ground fault current distribution when a ground fault occurs in HV substations located in an urban areas, Progress in Electromagnetic Research B 59 (2014) 167–179.

[6] Lj.M. Popović, Improved analytical expressions for the determination of the reduction factor of the feeding line consisting of tree single-core cables, European Transactions on Electrical Power 18 (2008) 190–203.
[7] R.J. Carson, Wave propagation in overhead wires with ground return, Bell Systems Technical Journal 5 (4) (1926) 539–554.
[8] F. Pollaczek, On the field produced by an infinitely long wire carrying alternating current, Electrische Nachrictentechnik III (9) (1926) 339–359.
[9] E. Sunde, Earth Conduction Effect in Transmission Systems, Van Nostrand Company, New York, 1949.
[10] J. Pohl, Influence of High-voltage Over-head Lines on Covered Pipelines, 1966. CIGRE, Paper No. 326.
[11] F. Dawalibi, A. Pinho, Computerized analysis of power systems and pipelines proximity effects, IEEE Transactions on Power Delivery 1 (2) (1986) 40–48.
[12] S.T. Sobral, W. Caustingheiras, V.S. Costa, D. Mukhedkar, Interference between faulted power circuits and communication circuits or pipelines – simplification using the decupled method, IEEE Transactions on Power Delivery 6 (4) (1991) 1599–1606.
[13] K.J. Satsios, D.P. Labridis, P.S. Dokopoulos, The influence of multi layer ground on electromagnetic field of an overhead power transmission line in the presence of buried conductors, Archiv fur Electrotechnik 80 (5) (1997) 343–348.
[14] G. Djogo, M.M.A. Salama, Calculation of inductive coupling from power lines to multiple pipelines and buried conductors, Electric Power System Research 41 (1) (1997) 75–84.
[15] G.C. Christoforidis, D.P. Labridis, P.S. Dokopoulos, Inductive influence on pipeline buried in multilayer soil due to magnetic fields from nearby faulted power lines, IEEE Transactions on Electromagnetic Compatibility 47 (2) (2005) 254–262.
[16] L. Bortels, J. Docaninck, J.I. Pineiro, C. Munteanu, V. Topa, A general applicable model for AC predictive and mitigation techniques for pipeline networks induced by HV power lines, IEEE Transactions on Power Delivery 21 (1) (2006) 210–217.
[17] D.A. Tsiamitros, G.K. Papagiannis, D.P. Labridis, P.S. Dokopoulos, Earth return path impedances of underground cables for the two-layers earth case, IEEE Transaction on Power Delivery 20 (3) (2005) 2174–2181.
[18] D.A. Tsiamitros, G.K. Papagiannis, P.S. Dokopoulos, Homogenous earth approximation of two-layer earth structures: an equivalent resistivity approach, IEEE Transactions on Power Delivery 22 (1) (2007) 658–666.
[19] CCITT Directives Concerning the Protection of Telecommunication Lines Again Harmful Effects from Electric and Electrified Railway Lines, ITU, Geneva, 1990 (Project report).
[20] Lj.M. Popović, New method for determination of inductive influence of one or more power lines on any of exposed metal installations, International Journal of Engineering Research and Application 6 (2) (2016) 36–46.
[21] Lj.M. Popović, Comparative analysis of grounding systems formed by MV cable lines with either uninsulated or insulated metal sheath(s), Electric Power System Research 81 (2) (2011) 393–399.
[22] IEEE Std 367-2012, Recommended Practice for Determining the Electric Power Station Ground Potential Rise and Induced Voltage from a Power Fault.
[23] NACE Standard Recommended Practice SPO177-2007, Mitigation of Alternating Current and Lightning Effects on Metallic Structure and Corrosion Control Systems, 2007.

FURTHER READING

[1] IEC 60364 Standard: Electrical Installations Buildings.

CHAPTER 8

Testing and Evaluating Grounding Systems in Urban Areas

8.1 INTRODUCTION

As is well known the relevant characteristics of the grounding system of a high-voltage (HV) substation at the design stage can be estimated only on the basis of calculations. However, because of a certain number of unavoidable approximations and idealizations of an actual physical model, only an approximate estimation can be achieved in this way. As a result, the final judgment concerning safety conditions within and in the vicinity of a certain HV substation can be obtained only by testing measurements and thereafter one can be sure that the substation will be fit for its intended purposes under all circumstances. However, in the case of HV/medium-voltage (MV) substations located in urban and suburban areas, these measurements usually do not involve the measurement of the actual value of the feeding line reduction factor [1—5].

Here, the presented method enables elimination of the influence of metal installations surrounding a test circuit on the accuracy of the results of measurements. For substations located in urban areas this influence is unavoidable and so intensive that it cannot be treated as negligible [1]. Consequently, the measured values of grounding system impedance and critical potential differences (the largest touch and step voltages) appearing on the grounding system can be significantly different in comparison with the actual ones. As a result, the final judgment regarding the achieved safety conditions of certain urban HV/MV substations can be completely incorrect. The presented method is developed on the basis of analysis of a complete electrical circuit formed during the ground fault. This circuit is formed not only by feeding line conductors but also by the many different metal installations involved with a fluctuating magnetic field around and along the feeding line.

According to contemporary practice in electric power distribution companies, the grounding system characteristics of an HV/MV substation

Practical Methods for Analysis and Design of HV Installation Grounding Systems
ISBN 978-0-12-814460-2
https://doi.org/10.1016/B978-0-12-814460-2.00008-8

are tested by using a version of the "fall-of-potential" method adapted for urban conditions. The basic measurement consists essentially of creating a test current through the grounding system of the tested substation via a phase conductor of the feeding line and the grounding system of the supply station or the adjacent transit substation supplied by the same line. In these substations the transformers that serve their own consumption are usually used as a voltage source for the test circuit formed in this manner. As is known, the potential measuring circuit is formed by using a provisory potential electrode positioned on the "remote earth." Here it is important to mention that the potential measuring circuit should be laid by following as much as possible the rules defined for the example in Ref. [6]. Then, the "remote earth" can usually be found at an approximately radial distance, which is between 300 and 500 m from the tested HV/MV substation. However, in areas with high values of soil resistivity this distance can be significantly greater [7,8].

Using the previously described procedure, it is possible to measure the potential and maximal potential gradients created by the test current on the grounding system of the tested HV/MV substation. To obtain the potential of the substation grounding electrode under actual ground fault conditions, the measured values are multiplied by the ratio of the actual ground fault current to the test current. If this potential is greater than the maximum tolerable touch voltages, then the potential differences appearing as a consequence of potential transferred to the nearest MV/LV substation should be examined. They are also determined under actual ground fault conditions by multiplying the test values with the ratio of the actual ground fault current to the test current. Thus it could be said that the value of the substation grounding impedance is unnecessary data in the procedure of the estimation of HV substation safety conditions.

The foregoing procedure, however, provides the real picture concerning the safety conditions inside and in the vicinity of a particular HV/MV substation, only if the test current circulates over the entire length of the feeding line or from the source to the tested substation (Section 3.2). A problem arises because this requirement is not always easy to fulfill in practice. Namely, in many cases there are one or more transit stations between the tested and source substation. This/these substation(s) should be temporarily out of operation and it can be very difficult to obtain permission to interrupt the power supply to so many consumers, including those that are essential for the normal functioning of public services in a city. As a rule the nearest transit substation is therefore used as an auxiliary grounding electrode in forming an indispensable test circuit. However, in these cases the metal installations participating as return paths of the test current (ground wire[s] in the case of an overhead line, cable metal sheaths in the case of a cable line, low-voltage [LV] network neutrals, and other underground metal installations such as water pipes and railway tracks) are not entirely the same as in the case of an actual ground fault. The current appearing during an actual ground fault circulates over the entire length of the feeding line and involves some additional metal installations participating as return paths for ground fault current. According to this, test current distribution generally cannot be the same as in the case of the actual ground fault current.

The described problem was first noted in Ref. [9] and considered in a simple example in Ref. [6]. Also numerical methods have been developed with the aim of solving this problem at the design stage [10,11]. They are based on the following procedure: grounding system elements, including an HV feeding cable line, are fictitiously divided into several sections to model their grounding effects and mutual inductive influence. However, the problems appear already at the stage of preparing necessary input data for their application. The structure and topology of the underground metal installations surrounding the feeding line are usually uncertain or somewhere completely unknown. Also data concerning equivalent resistivity of the soil along the feeding line are quite uncertain. Thus by using these methods it is not possible to obtain a required grounding system evaluation, even if relying on the possibilities offered by modern computer systems.

The possibility of determining actual ground fault current distribution in the grounding system of an urban HV/MV substation is purposely investigated, and the obtained results are presented in Refs. [1–3]. On the basis of these mainly experimental investigations, and on the basis of introducing a fictitious equivalent conductor (later on referred to as equivalent conductor), a corresponding analytical procedure is developed for determining the actual reduction factor of the feeding line passing through the urban area. The relevant parameters of the mentioned equivalent conductor are determined under the condition that its influence on ground fault distribution is equal to the one obtained by site measurements for the influence of all metal installations actually existing around and along the entire feeding line (Section 3.2).

As has already been shown (Section 3.4) the portion of the ground fault current flowing through the earth in the considered cases is several times smaller than was previously known. This fact throws a completely new light on the entire grounding problem of urban substations and dramatically changes our previous perception of the magnitude of this problem. It can now be seen in a realistic framework that generally leads to more economical solutions. Certainly, greater economic effects can be expected in cases where, because of high soil resistivity and/or a high short-circuit level, special measurements (e.g., bare copper wire laid in the same trench as the cable feeding line, counterpoises, etc.) have been foreseen as necessary [12–14].

The test method presented here is based on the results of the investigations presented in Refs. [1–4]. The developed procedure enables elimination of the influence of metal installations surrounding the test circuit on the result of measurements of the grounding system impedance. The input data consist of measured values of the test current and its fraction appearing in the neutral conductor (metal sheath in the case of a cable line or ground wire in the case of an overhead line) of the feeding line. On the basis of correctly determined grounding system impedance of the tested substation, the proposed preliminary estimation procedure of safety conditions can be sufficient for correct final judgment about safety conditions in many practical cases. By application of the presented method the need for undesirable and harmful interruptions in the supply of electrical energy from the transit substation is avoided.

Finally, it is most important to mention that, without using the presented method, the assessment of the safety conditions within and in the vicinity of a certain distribution HV/MV substation can be completely wrong. In the other words, the prescribed safety requirements could be considered as fulfilled where this is not the case.

8.2 DESCRIPTION OF THE BASIC PROBLEM

As already said, all safety problems of an HV substation emanate from one fraction of the total ground fault current. This is the current that is dissipated through the substation grounding system into the surrounding earth. Thus for correct estimation of safety conditions it is of prime importance to correctly estimate this fraction of the ground fault current. However, this is not an easy task in the case of substations located in urban and suburban areas. Grounding systems of HV distribution substations consist of a substation grounding electrode and many outgoing MV cable lines acting as external grounding electrodes and/or conductive connections with the grounding systems of the supplied MV/LV substations. Outgoing cable lines spontaneously form a very complex grounding system on a large urban area around these substations. This grounding system includes, through the terra-neutral (TN) earthing system in the LV network and consumer installations, many known and unknown metallic installations typical for the urban area [15,16]. As a consequence, outgoing cable lines simultaneously become the conductive connections with the metal installations laid along the same street(s) as the feeding line. Due to this, the current appearing in the metal sheath of each outgoing cable line depends on its disposition in relation to the other elements of the grounding system and in relation to the feeding line. Thus it is not difficult to imagine that the current passing through the grounding system elements only as a consequence of the inductive influence of the feeding line is distributed in a very complex way between the many external elements of the grounding system. Obviously, the fault current passing through the elements of the grounding system has two, essentially different, fractions. One of them is dissipated into the surrounding earth, while the other is induced in the metal installations surrounding the feeding line. Since the split between these two fractions occurs along the many grounding system elements and belowground, none of these fractions can be separated and determined by direct or indirect measurements. However, without the correctly determined current dissipated into the earth, it is not possible to correctly determine the grounding system impedance of the tested substation. The problem is solved by developing the test procedure presented in Ref. [4]. This enables the determination of current dissipating into the earth by using measured values of the test current and the current induced in one of the neutral conductors of the feeding line or its section used for the test circuit.

8.3 GROUNDING SYSTEM IMPEDANCE DETERMINATION

8.3.1 Test Circuit Formed Through the Cable Line

Let us consider the case where the test circuit is formed by using the feeding line section between the tested and the nearest transit substation serving as an auxiliary current electrode. As first we will assume the case where this line section is realized using three single-core cables.

In practical conditions, besides the applied phase conductor and its associated metal sheath, many other metal installations typical for urban areas spontaneously participate in forming the test circuit. Many of these installations are laid in the same sidewalk(s) or in the same street(s) as a feeding line section used for obtaining the test circuit. Also many of these installations are mutually conductively connected by applying the TN earthing system in each building along the contemporary town streets. Thus independent of their basic functions, these installations spontaneously form a large and unique network of different mutually connected conductors. Its structure and topology are usually very complex, depend on many local factors, and cannot be visually determined or verified. Finally, it can be said that the test circuit in a typical urban area represents a very complicated electrical circuit placed mainly under the surface of the ground or in places that usually are not accessible for measurements.

To simplify the problem and enable its solution, all known and unknown metal installations participating in test current underground distribution are substituted by an equivalent conductor. This is cylindrically shaped and can be treated as an additional neutral line conductor. The simplified physical model of the test circuit obtained on the basis of this substitution is presented as shown in Fig. 8.1.

The used notation has the following meaning:

- *A*(*B*)—transit (tested) substation,
- U_a—auxiliary voltage source,
- I_t—test current,
- I_s—current induced in the cable metal sheath,
- I_c—current induced in the introduced equivalent conductor,
- I_e—current injected into the earth through the grounding system of substation *B*,
- Z_a (Z_b)—grounding system impedance of substation *A*(*B*), and
- *G*—remote ground.

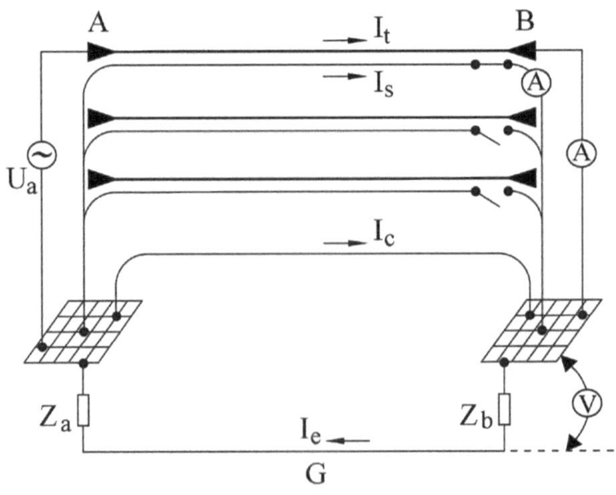

Figure 8.1 Simplified model of the measurement circuit [4].

For further circuit simplification, the test circuit should be formed in the following manner: only the metal sheath of the cable with the test current should be grounded at both of the test circuit ends, while the other two metal sheaths should be ungrounded at least at one of the test circuit ends as shown in Fig. 8.1.

Then by using the simplified electrical circuit substituting physical model in Fig. 8.1, any actual test current distribution can be presented by the equivalent circuit shown in Fig. 8.2.

The used notation has the following meaning:

- Z_s—self-impedance of the cable metal sheath (Ω),
- Z_c—self-impedance of the introduced equivalent conductor (Ω),
- Z_{0s}—mutual impedances between the phase conductor and its metal sheath with earth as a common return path (Ω),
- Z_{0c}—mutual impedance between the phase conductor and introduced equivalent conductor with earth as a common return path (Ω), and
- Z_{sc}—mutual impedance between the cable sheath and introduced equivalent conductor with earth as a common return path (Ω).

To determine grounding impedance Z_b by measurements, it is necessary to have test values of the potential appearing at the substation grounding electrode and current I_e (Fig. 8.1). However, current I_e is dissipated into the ground through the many elements of the grounding system and cannot be separated from current I_c that also, as mentioned earlier, passes through the elements of the grounding system and cannot be separated for direct measurement. Due to this, an actual value of the grounding system impedance cannot be obtained by direct use of test results. Its value, according to the former practice, is obtained as a ratio between the tested grounding system (electrode) potential and the difference of the measured currents I_t and I_s. However, according to Fig. 8.2, in this way it is ignored

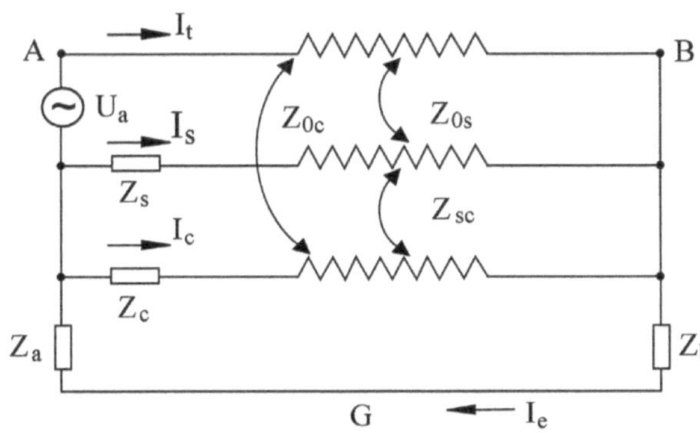

Figure 8.2 Simplified equivalent circuit [4].

by the current passing through the surrounding metal installations. As a result, this impedance is just an apparent impedance that can be, as presented later in Section 8.4, several times smaller than the actual one.

On the basis of the equivalent circuit in Fig. 8.2 for the determination of I_e, it is a prerequisite that the relevant parameters of the introduced equivalent conductor be determined. The cross-sectional shape and spatial position of this conductor toward the used feeding cable line have been, according to Section 3.2, defined in the manner shown by Fig. 8.3.

The used notation has the following meaning:

- r_e—medium radius of the cylinder representing an equivalent conductor.

On the basis of the imagined physical appearance of the introduced equivalent neutral conductor (Figs. 8.1 and 8.3) the mathematical expressions necessary for the determination of all impedances of the equivalent circuit in Fig. 8.2 become known. These impedances are, according to Ref. [15], given by:

$$Z_s' = R_S' + \omega_t \frac{\mu_0}{8} + j\omega_t \frac{\mu_0}{2\pi} \ln \frac{\delta_t}{r_{sh}}, \; \Omega/\text{km}, \tag{8.1}$$

$$Z_{0s}' = \omega_t \frac{\mu_0}{8} + j\omega_t \frac{\mu_0}{2\pi} \ln \frac{\delta_t}{r_{sh}}, \Omega/\text{km}, \tag{8.2}$$

$$Z_c' = R_e' + \omega_t \frac{\mu_0}{8} + j\omega_t \frac{\mu_0}{2\pi} \ln \frac{\delta_t}{r_e}, \; \Omega/\text{km}, \tag{8.3}$$

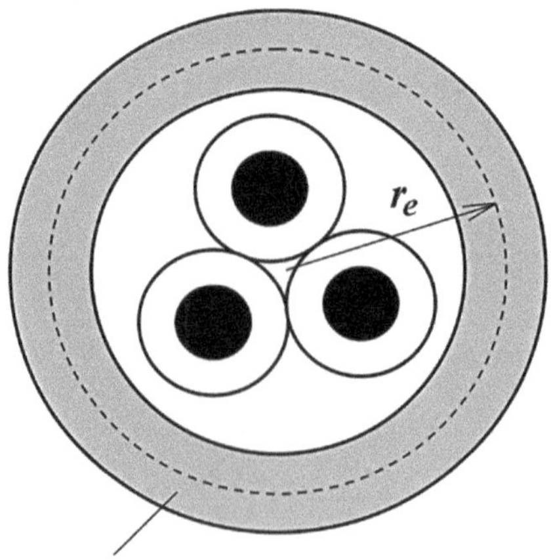

Figure 8.3 Equivalent conductor and three single-core cable lines [4].

$$Z'_{0c} \approx Z'_{sc} = \omega_t \frac{\mu_0}{8} + j\omega_t \frac{\mu_0}{2\pi} \ln\frac{\delta_t}{r_e}, \Omega/\text{km}, \tag{8.4}$$

where R'_S is the cable sheath resistance per unit length (Ω/km), R'_e is the equivalent conductor resistance per unit length (Ω/km), r_{sh} is the medium radius of the cable sheath (m), ω_t is the angular test frequency, $\omega_t = 2\pi f t$, μ_0 is the magnetic permeability of a vacuum, $4\pi \times 10^{-7}$ Vs/Am, μ_r is the relative magnetic permeability, δ_t is the equivalent earth penetration depth (m), and prime ($'$) denotes quantities given per unit length.

The equivalent earth penetration depth is determined by:

$$\delta_t = 658\sqrt{\frac{\rho}{f_t}}, \ m, \tag{8.5}$$

where ρ is the equivalent soil resistivity along and around the cable line (Ωm) and f_t is the test circuit frequency (Hz).

On the basis of the given expressions, it can be seen that for determining impedances Z_c, Z_{0c}, and Z_{sc}, it is a prerequisite that the relevant parameters of the equivalent conductor be known. This conductor can also be considered as an additional metal sheath for all three single-core cables of the feeding line (Fig. 8.3). By using the equivalent circuit in Fig. 8.2 and the assumption that grounding impedances Z_a and Z_b are negligible ($Z_a \approx Z_b \approx 0$), necessary parameters R'_e and r_e can be determined on the basis of the following equation:

$$r_m Z'^2_{sc} - Z'_c\left[Z'_s(r_m - 1) + Z'_{0s}\right] = 0. \tag{8.6}$$

In the given equation, r_m represents a coefficient expressing reduction of the current I_e caused by the current induced only in cable sheath I_s. It is determined by:

$$r_m = 1 - \frac{I_s}{I_t}, \tag{8.7}$$

where currents I_t and I_s are, according to Fig. 8.1, obtained by measurements.

Since Eq. (8.6) gives the relationship between complex quantities, it can be presented as the following system of two equations:

$$\begin{aligned} \text{Re}\left\{r_m Z'^2_{sc} - Z'_c\left[Z'_s(r_m - 1) + Z'_{0s}\right]\right\} = 0 \\ \text{Im}\left\{r_m Z'^2_{sc} - Z'_c\left[Z'_s(r_m - 1) + Z'_{0s}\right]\right\} = 0. \end{aligned} \tag{8.8}$$

On the basis of this system of equations, it is possible to determine the relevant parameters of the introduced equivalent conductor in each concrete case. Having in mind the assumptions made by derivation of this system of equations ($Z_a \approx Z_b \approx 0$), it is necessary to mention that, for its application, the following inequality should be satisfied:

$$Z_a \approx Z_b \ll Z_s' L, \tag{8.9}$$

where L is the length of the feeding line section between stations A and B (km).

In other words, a current appearing in the metal sheath only as a consequence of the potentials on the grounding electrodes of substations A and B should be negligible. In actual conditions, quantitative relationships between relevant parameters are such that this criterion is satisfied for almost each substation and its feeding line section in an HV cable network. Simultaneously, it means that the influence of the length of the feeding line section used for a test circuit on test current distribution can be considered as negligible.

When data regarding parameters R_c' and r_e are determined, the impedances Z_c, Z_{0c}, and Z_{sc} can be determined by using Eqs. (8.3) and (8.4). Then, by using the equivalent circuit given in Fig. 8.2 and Kirchhoff's rules, the following equation can be written:

$$I_e = \left(Z_s' Z_c' - Z_{0s}' Z_c' - Z_{0c}' Z_s' - Z_{sc}' Z_{0c}'\right) I_t - \left(Z_{sc}' Z_s' - Z_{sc}'^2\right) I_s. \tag{8.10}$$

Finally, the grounding system impedance of the tested substation B can be, according to Fig. 8.1, determined by:

$$Z_b = \frac{V_t}{I_e}, \tag{8.11}$$

where V_t is the potential of the grounding electrode of substation B (V).

With the aim of eliminating the influences of the test circuit and metal installations surrounding the potential measurement circuit, this potential is, according to Ref. [13], determined by:

$$V_t = k_1 k_2 V_m, \tag{8.12}$$

where V_m is the tested electrode potential obtained by measurement (V), k_1 is the voltmeter correction factor, and k_2 is the relative position correction factor.

Voltmeter correction factor is given by:

$$k_1 = \frac{R_V + (Z_p + R_E)}{R_V}, \tag{8.13}$$

where R_V is the internal resistance of the voltmeter (nameplate value), $Z_p + R_E$ is the impedance of the potential measurement circuit seen from the voltmeter terminals (this value is measured by connecting a Megger directly to the voltmeter terminals), and R_E is the grounding resistance of the potential electrode.

Relative position correction factor k_2 is determined by:

$$k_2 = \frac{1}{\frac{p}{d_a} + \frac{p}{d_b} + \frac{p}{d_c} + 1}, \tag{8.14}$$

where $p = \frac{1}{2}\sqrt{\frac{A_B}{\pi}}$, m, A_B is the area of the substation under measurement (m^2), d_a is the distance between the current and potential electrodes (m), d_b are the distances between the potential electrode and the limits of the tested electrode (m), and d_c is the distance between the current electrode and the limits of the tested electrode (m).

The value of the grounding resistance R_E should be kept small to improve the quality of the measurements. However, when an electronic voltmeter of high internal resistance ($>10^7\ \Omega$) is used, this requirement is of secondary importance. Also the direction of the cable used for this potential measurement should be as perpendicular as possible to the current test circuit to minimize interference.

8.3.2 Test Circuit Formed Through the Overhead Line

Overhead lines are also used in contemporary HV distribution networks, especially in suburban areas, or in areas with a lower degree of urbanization. As a result, when the feeding line section between the tested and transit substations is an overhead line, the test current underground distribution should also be considered from the standpoint of the inductive influence of metal installations surrounding the test current circuit.

The only difference in comparison with the previous case lies in fact that, instead of the phase conductor and the metal sheath(s) of the cable line, we have here a phase conductor and ground wire(s) of an overhead line. Since the analogy is obvious, the same physical model and the same equivalent circuit can be used as in the previous case (Figs. 8.1 and 8.2), only instead of Z_s, Z_{0s}, Z_{0c}, and Z_{sc} used in Fig. 8.2 we now have, seen respectively, the following notations:

- Z_w—self-impedance of the ground wire(s) (Ω/km),
- Z_{0w}—mutual impedance between the ground wire(s) and the most distant (critical) phase conductor (Ω),

- Z_{0c}—mutual impedance between the mentioned phase conductor and the introduced equivalent conductor (Ω), and
- Z_{wc}—mutual impedance between ground wire(s) and the introduced equivalent conductor (Ω).

The impedances Z_w and Z_{0w} given per unit length are, according to Ref. [17], determined by:

$$Z'_w = R'_w + \omega\frac{\mu_0}{8} + j\omega\frac{\mu_0}{2}\left(\frac{\mu_r}{4} + \ln\frac{\delta}{r_w}\right), \tag{8.15}$$

$$Z'_{0w} = \omega\frac{\mu_0}{8} + j\omega\frac{\mu_0}{2}\left(\frac{\mu_r}{4} + \ln\frac{\delta}{d_w}\right), \tag{8.16}$$

where R'_w is the resistance of the ground wire(s) per unit length (Ω), r_w is the equivalent ground wire radius for one or more ground wires (m), and d_w is the distance between the ground wire(s) and the most distant (critical) phase conductor (m).

Since we are interested only in the critical cases from a safety condition standpoint, the phase conductor most distant from the ground wire(s) is used for the test circuit.

In this case we will assume that the equivalent conductor is also a cylinder and that its center is in the middle of the fictitious line connecting the ground wire and mentioned phase conductor. On the basis of these assumptions, one can obtain the relevant parameters of the introduced equivalent conductor and the grounding system impedance of substation B by using the same calculation procedure as in the case of the cable feeding line. However, it is important to emphasize here that the part of the test current appearing in the ground wire(s) as a consequence of the potentials on the grounding electrodes at the test circuit ends can be significant in the cases of these lines. As a result, the current induced in the ground wire(s) should be measured at a place that is sufficiently distant from substation B. This distance can be, according to Eq. (2.80), approximately determined by the active length of the ground wire(s), or:

$$N_a = \frac{3}{\ln\left(1 + \sqrt{\frac{\left|L_S Z'_w\right|}{R}}\right)}, \tag{8.17}$$

where N_a is the active length of the ground wire(s) as grounding conductor, expressed in number of average spans, R is the average tower footing resistance (Ω), and L_S is the length of one span of average length (km).

In cases where the length of the section between substations A and B is shorter than two active ground wire lengths, measurement of the induced current should be performed somewhere in the middle of this feeding line section.

8.3.3 Preliminary Estimation of Safety Conditions

On the basis of the determined value of the grounding system impedance and the value of the anticipated maximal ground fault current in substation B, the potential on the grounding electrode of this substation can be determined by:

$$V_g = r_c Z_b I_F, \tag{8.18}$$

while the potential differences (touch and step voltages) essential for the estimation of the safety conditions can be determined by:

$$U_m = U_{mt} \frac{V_g}{V_t}, \tag{8.19}$$

where I_F is the anticipated maximal ground fault current passing through the grounded neutral points in supply network (A), r_c is the feeding line reduction factor obtained by calculation, U_m is the maximal touch voltage that can appear on the grounding system (V), and U_{mt} is the maximal touch voltage obtained by test measurements (V).

The reduction factor of the feeding line of three single-core cables is, according to Section 3.1, calculated as follows:

- For cables laid in flat formation:

$$r_c = \frac{R'_S}{Z' + Z'_m + Z'_{m2} - Z'_m \dfrac{Z'_m - Z'_{m2}}{Z' - Z'_m}}, \tag{8.20}$$

- For cables laid in three-angular formation:

$$r_c = \frac{R'_S}{R'_S + 3\omega \dfrac{\mu_0}{8} + j3\omega \dfrac{\mu_0}{2\pi} \ln \dfrac{\delta}{\sqrt[3]{r_{sh} d^2}}}, \tag{8.21}$$

where r_c is the calculated feeding line reduction factor, Z'_m is the mutual impedance between two adjacent cable sheaths with earth as common return path (Ω), Z'_{m2} is the mutual impedance between two outer cable sheaths with earth as common return path (Ω), and d is the distance between two adjacent cables.

Mutual impedance Z'_m is, according to Ref. [15], determined by:

$$Z'_m = \omega\frac{\mu_0}{8} + j\omega\frac{\mu_0}{2\pi}\ln\frac{\delta}{d}, \ \Omega/\text{km}. \tag{8.22}$$

Mutual impedance Z'_{m2} is also determined by Eq. (8.22), only, according to Fig. 3.4, "2*d*" appears in it instead of "*d*."

On the basis of Fig. 3.5, it is not difficult to show that the reduction factor for the cable line with the applied cross-bonding technique can be calculated by:

$$r_c = \frac{2r_o + r_i}{3}, \tag{8.23}$$

where r_o is the reduction factor when ground fault current circulates through the cable taking one of the outer positions and r_i is the reduction factor when ground fault current circulates through the cable taking the inner position.

Reduction factor r_o is calculated by Eq. (8.20), while reduction factor r_i is, according to Section 3.1, calculated by:

$$r_i = \frac{R'_S}{Z' + 2Z'_m - Z'_m\dfrac{Z'_m - Z'_{m2}}{Z' - Z'_m}}. \tag{8.24}$$

However, because of the influence of the metal installations surrounding a feeding line, the actual value of the reduction factor can be significantly smaller than the one obtained by calculations, as can be seen in Section 3.4. Due to this, the previously described estimation procedure gives results that are excessively on the conservative side for substations located in typical urban surroundings.

When the test circuit passes through an area with an obviously greater degree of urbanization than the rest of the feeding line, one other preliminary estimation can be made. Instead of using the calculated reduction factor, one can use the actual reduction factor of the feeding line already obtained for the transit substation. Certainly, this reduction factor can be used at the design stage of substation *B*, as it is more favorable than the one obtained by calculation.

In critical cases, when the preliminary determined maximal transferred potential and touch voltage exceed tolerable limits, it is necessary to make an estimation by using the actual value of the reduction factor. This means that the test circuit should be formed in such a manner that the test current circulates over the entire feeding line length. Only in this manner can the actual contributions of all surrounding metal installations participating in ground fault current distribution be taken into account [4].

It should also be mentioned that the HV/MV distribution substations usually have more than one connected HV line that can be used as a feeding line in certain regular or irregular supplying circumstances in the HV network. The maximum (critical) fault current dissipated into the earth in such cases cannot be known in advance, i.e., only on the basis of the largest possible value of the total ground fault current.

The largest value of so-called earth current can be in cases where total ground fault current, I_F, is not the largest. This can be encountered in cases where the feeding line causing maximum total ground fault current passes through the most urbanized area. Due to this, in some cases it is necessary to perform test measurements for each of the possible feeding lines, and, on the basis of comparing the different values for earth current, obtain the maximum possible value of this current, or the worst case from the standpoint of safety conditions.

According to the presented analytical procedure, the application of this method takes into account the influence of all relevant factors and parameters, including those whose influence is negligibly small. This means that this method enables correct estimation for each, including extremely complex, practical situation. Some inaccuracy can appear only as a consequence of the inductive influence of the nearby power distribution lines. This problem can be efficiently overcome by using for the test current a somewhat higher frequency that can be easily discriminated from the omnipresent power frequency currents.

8.4 PRACTICAL EXAMPLE

Experimental investigations on the influence of metal installations typical of urban areas with respect to the value of the feeding line reduction factor are performed on a cable line that supplies, in series, two substations in the 110 kV distribution network in Beograd, Serbia [1]. The length of the line to the closer of the supplied substations measured from the supply substation is 2320 m, while the total feeding line length to the tested substation is 6590 m. The line is realized by single-core cables, laid in a triangular formation on the entire length. The cross-bounding technique used for reduction of currents circulating through the cable sheaths was not applied to this line.

In the areas through which the line passes, equivalent soil resistivity is roughly estimated to be in the range from 30 to 50 Ωm, and being less favorable from the standpoint of ground fault current distribution the lower of these values is used for necessary calculations. The line section between the source substation and the transit one passes through an area with a lower degree of urbanization than the rest of the line. The phase conductors are made of aluminum with a cross-section of 1000 mm^2, while the metallic sheaths are made of copper strings with a total cross-section of 95 mm^2 and a medium diameter equal to 91 mm. The main elements of the grounding system in both of the supplied substations are the substation building foundation and the 44 outgoing cable lines realized by an armored lead-sheathed three-core cable.

In the first attempt, the test circuit was formed by using only a feeding line section between the tested and the transit substations. The length of this section is 4270 m. The second test circuit was formed by using the whole feeding line. While measurements were being taken in both of these lines, only the metal sheath of the cable with the test current was grounded at both of the test circuit ends.

On the basis of the measurements of the test current and current through the cable sheath, the following results have been obtained for the coefficient defined by Eq. (8.7):

$$r_m = 0.2555 + j0.3790, \quad \text{for the shorter test circuit,}$$

$$r_m = 0.2415 + j0.3623, \quad \text{for the longer test circuit.}$$

With the aim of enlarging the ground current, I_e, and obtaining the apparent value of the ground system impedance, an additional measurement was performed. This measurement was made by disconnecting all feeding line metal sheaths at one of the test circuit ends. Then, the test current and potential at the grounding electrode of the tested 110/10 kV substation were measured. The results obtained were:

$$I_t = 136\text{ A and } V_t = 1.14\text{ V}, \quad \text{for shorter test circuit,}$$

$$I_t = 85\text{ A and } V_t = 1.20\text{ V}, \quad \text{for longer test circuit.}$$

On the basis of Eq. (8.8) and the measurement results obtained for r_m, it follows that the relevant parameters of the equivalent conductors are:

$$R'_e = 0.031\ \Omega/\text{km and } r_e = 7.1\text{m}, \quad \text{for the shorter test circuit,}$$

$$R'_e = 0.053\ \Omega/\text{km and } r_e = 8.1\text{m}, \quad \text{for the longer test circuit.}$$

By using Eq. (8.10) and results obtained for R'_e and r_e, the values obtained for the earth current are:

$$I_e = (4.23 - j14.15)\text{ A}, \quad \text{for the shorter test circuit,}$$

$$I_e = (5.89 - j15.00)\text{ A}, \quad \text{for the longer test circuit.}$$

When this current is known, on the basis of the measured value of the corresponding V_t, the identical value for the substation grounding impedance is obtained in both of the cases being considered. Certainly, there can only be one actual value of the grounding system impedance and it is:

$$Z_b \approx 0.075\ \Omega.$$

Without taking into account the influence of the metal installations surrounding the test circuit, this impedance would be considered as 70.9% smaller for the longer and 88.6% smaller for the shorter test circuit. In the previous investigations presented in Ref. [1] these apparent values were treated as actual ones. This is probably the main explanation of the fact that the measured values of the grounding system impedance for all 110/10 kV substations in the distribution network in Beograd were almost always several times smaller than the ones obtained by calculations at the design stage.

On the basis of the determined actual value of the grounding system impedance and the value of the calculated reduction factor of the feeding line, preliminary estimation of the safety conditions for the tested HV/MV substation becomes possible.

By using Eq. (8.21) the calculated reduction factor in the case of the considered line is:

$$r_c = 0.032 - j0.105.$$

However, since the feeding line section from the source substation to the transit station passes through an obviously less urbanized area, more accurate preliminary estimation can be made by using the actual reduction factor of the feeding line for the transit substation. By using the given relevant data, this reduction factor is:

$$r_t = 0.041 - j0.036.$$

Then, the final judgment of the safety conditions realized in the tested substation can be obtained by using the value of the actual feeding line reduction factor. This value in the case under consideration is:

$$r_a = 0.030 - j0.021.$$

To obtain a final estimation, a necessary condition is that the test current circulates over the entire feeding line length. However, fulfillment of this condition, as mentioned earlier, requires transit substation(s) to be temporarily out of operation.

The value of the reduction factor for the feeding line section between the tested and transit substations in the considered case is only 0.019–j0.010. Since this value is significantly lower than the actual reduction factor for the considered feeding line, it is clear that the test circuit formed by using the transit substation grounding system as an auxiliary current electrode in this case can give inaccurate estimation of the achieved safety conditions. Thus it can be used only for obtaining the actual value of the grounding system impedance and, on this basis, the correct preliminary estimation of the safety conditions.

The presented results relate to only two different cases from the standpoint of the feeding line length and the surrounding metal installations (the feeding line to the tested substation and the feeding line to the transit substation). However, they show themselves as significant for all cases when, for test purposes, only the nearest section of the entire feeding line is available.

On the basis of the performed experimental investigations and the results presented here, for current engineering practice it is important to emphasize the following observations. The grounding problem of distribution HV/MV substations has been efficiently and cost effectively solved, thanks to the different metal installations that exist in urban areas, independent of our need for solving this problem. Some of them, surrounding these substations, significantly reduce the impedance of their grounding systems [7,10,12,18], while others, surrounding their feeding lines, according to the results presented in Refs. [1−4] and here, significantly reduce the fraction of the ground fault current dissipating into the surrounding earth.

Not to rely only on a single practical example, we will assume the cases where the urban environment is identical, but only the equivalent soil resistivity is gradually increased. Then, as well as in the case of the current through feeding line neutral conductors, the current through the surrounding metal installations becomes greater and greater, while the current through the earth becomes smaller and smaller. According to this, the final calculation results will be slightly on the conservative side if we adopt a value of the equivalent soil resistivity that is considered as minimal as possible for the urban area along a considered feeding line. The effects of the reduction of the current dissipating into the surrounding earth are also greater if the urban environment is on the greater degree of urbanization.

Then, on the basis of the presented results, it is clear that a significant part of the test current passes through the different metal installations surrounding the feeding line. This means that under actual ground fault conditions, this current can be several kiloamperes. Consequently, from the standpoint of public safety it is important to stress that all these metal installations should be mutually connected. This requirement, as is well known, should already be fulfilled through the consistent application of TN earthing systems in LV networks and customers' installations.

The test method presented here enables a correct preliminary estimation and final judgment regarding safety conditions within and in the vicinity of HV/MV substations supplied through one or more transit substations. The influence of the metal installations surrounding the test circuit is taken into account through the measured values of the test current and the current induced in one of the neutral conductors of the feeding line section used for the test circuit.

Finally, since it is quite a realistic assumption that similar urban conditions for all other HV/MV substations can be anticipated all over the world, one can say that with the presented testing method, the methodology for solving the grounding problem of HV substations located in urban areas is rounded. Bearing in mind all presented facts the method could be used as a foundation for revision of the current versions of the corresponding technical standards, e.g., Ref. [19].

REFERENCES

[1] Lj.M. Popović, Influence of the metallic installations surrounding the feeding cable line on the ground fault current distribution in supplied substations, IEEE Transactions on Power Delivery 23 (4) (2008) 2583–2590.

[2] Lj.M. Popović, Ground fault current distribution when ground fault occurs in HV substations located in an urban area, Progress in Electromagnetic Research B 59 (2014) 167–179.

[3] Lj.M. Popović, Determination of the actual reduction factor of HV and MV cable line passing through urban and suburban area, FACTA UNIVERSITATIS-Series: Electronics and Energetics 27 (1) (2014) 25–39.

[4] Lj.M. Popović, Testing and evaluating grounding systems for substations located in urban areas, IET Generation, Transmission and Distribution 5 (2) (2011) 231–238.

[5] Lj.M. Popović, Actual Parameters of Power Lines Passing through Urban Areas, LAP LAMBERT Academic Publishing, 2015.

[6] P.R. Pilai, E.P. Dick, A review on testing and evaluating substation grounding systems, IEEE Transactions on Power Delivery 7 (1) (1992) 53–61.

[7] Lj.M. Popović, Comparative analysis of grounding systems formed by MV cable lines with either uninsulated or insulated metal sheath(s), Electric Power System Research 81 (2) (2011) 393–399.

[8] Lj.M. Popović, Active length of cable lines with uncoated metal sheaths as long grounding conductors, European Transaction on Electrical Power 22 (2) (2012) 206–215.

[9] J. Fortin, H.G. Sarmieto, D. Mukhedkar, Field measurements of ground fault current distribution at LG – 2, Quebec, IEEE Transactions on Power Delivery 1 (3) (1986) 48–60.

[10] S.A.P. Meliopoulos, An advanced computer model for grounding system analysis, IEEE Transactions on Power Delivery 8 (1) (1993) 13–23.
[11] J.M. Nahman, V.B. Djordjevic, D.D. Salamon, Grounding effects of HV and NV underground cables associated with urban distribution substations, IEEE Transactions on Power Delivery 17 (1) (2002) 111–116.
[12] J. Wilas, D. Mukhedkar, V. Fernandes, A. Magalhaes, Grounding grid design of a transition station – a typical example of fault transfer, IEEE Transactions on Power Delivery 5 (1) (1989) 124–129.
[13] S.T. Sobral, J.J. Horta Santos, D. Mukhedkar, A proposal for ground measurement techniques in substations fed exclusively by power cables, IEEE Transactions on Power Delivery 3 (4) (1988) 1403–1409.
[14] J. Waes, M. Riet, F. Proroost, S. Cobben, Measurements of the current distribution near a substation during a single phase to ground fault, in: Proc. CIRED Conference, 2003.
[15] IEC 60364 Standard: Electrical Installations Buildings.
[16] Int. Std., CEI IEC 61936 – 1 Power Installations Exceeding 1kV A.C. Part 1: Common Rules, First Edition 2002–10.
[17] Int. Std., Ref. CEI/IEC 60908-3, Ed. 0.3 2009-3, Short-circuits Currents in Three Phase A.C. Systems – Part 3: Currents During Two Separate Simultaneous Line-to-earth Short Circuits and Partial Short – Circuits Following Through Earth.
[18] Lj.M. Popović, Practical method for the analysis of earthling systems with long external electrodes, IEE Proceedings C 140 (3) (1993) 213–220.
[19] IEEE, Guide for Safety in Substations Grounding, Std. 80 – 2000.

INDEX

Note: 'Page numbers followed by "f" indicate figures and "t" indicate tables.'

H